GERMANY'S TIGER TANKS
VK45.02 TO TIGER II

Also by Thomas L. Jentz

GERMANY'S TIGER TANKS - TIGER I & II: COMBAT TACTICS

GERMANY'S PANTHER TANK: THE QUEST FOR COMBAT SUPREMACY

PANZERTRUPPEN: 1933-1942

PANZERTRUPPEN: 1943-1945

GERMANY'S TIGER TANKS

VK45.02 TO TIGER II

Thomas L. Jentz & Hilary L. Doyle

Schiffer Military History
Atglen, PA

Book Design by Robert Biondi.

Library of Congress Catalog Number: 97-67431.

Printed in the United States of America.
ISBN: 0-7643-0224-8

We are interested in hearing from authors with book ideas on related topics.

Published by Schiffer Publishing Ltd.
4880 Lower Valley Road
Atglen, PA 19310
Phone: (610) 593-1777
FAX: (610) 593-2002
Email: Schifferbk@aol.com.
Please write for a free catalog.
This book may be purchased from the publisher.
Please include $3.95 postage.
Try your bookstore first.

Contents

Introduction

As data were gathered for this book, it was astounding that so much new information was found that had never been published before on such a well-known Panzer. What really amazed Hilary and myself were the new discoveries every time we looked in detail at a Tiger in a museum or thoroughly studied the manuals. We ourselves hadn't recognized some key modifications before, such as three different commander's cupolas and three different loader's hatches on Tiger II turrets. In addition, we found that the shape of the **Serien-Turm** (series turret) for the Tiger II is asymmetrical. The left side of the **Serien-Turm** is bowed outward 2 cm wider than the right side. Our research proved that no one had ever accurately created scale drawings of an entire Tiger II, not even in the original drawings from Henschel. To ensure full coverage of the modifications, extra effort went into obtaining and including rarely seen top and internal views of the surviving Tiger II.

Over twenty years of intensive research went into finding the original documents needed to create this history of the development, characteristics, and tactical capabilities of the Tiger II. An exhaustive search was made for surviving records of the design/assembly firms (including Krupp, Henschel, Porsche, and Wegmann), the **Heereswaffenamt**, the **Generalinspekteur der Panzertruppen**, the D656 series of manuals on the Tiger, and the war diaries with their supporting reports from German army units. This was supplemented by our collecting hundreds of photos and climbing over, under, around, and through almost every surviving Tiger II.

All information contained in this book is extracted directly from original source material. Not one single bit of information has been derived from other published books. If I couldn't find it in an original document, photograph, or surviving Tiger II, it isn't in this book. My sources of information are stated in the text. Those desiring footnotes and an extensive bibliography are advised to look elsewhere. My goal is to create an accurate record of events, not justify opinions derived from other publications.

Events only occur in one way. A clear picture of these events cannot be obtained from stories, military intelligence reports, expert opinion, analyses of raw data, or secondary sources. Either the events were recorded by a participant or they weren't. No amount of conjecture, supposition, reasoning, or opinion can fill gaps in the original records. Even though participants in the design of the Panzers made occasional mistakes in their records and sometimes slanted reports to be more or less favorable, they were the only direct observers. Any gross errors that may have made in a original report can be discovered by careful examination of the end product and comparison to other independently created original reports and drawings.

Many errors have crept into postwar publications through the use of popular nicknames for the German Panzers. If the author doesn't even know the correct name for a Panzer, what else is wrong with his information? The most commonly used names are often the most misleading. There never was a "Porsche" turret or a "Henschel" turret. Porsche and Henschel did not design turrets. Krupp designed the turrets. Porsche and Henschel designed the chassis upon which turrets were mounted. Even using correct nomenclature but sloppily mixed together, such as Tiger II Ausf.B, leaves people wondering just what was the **Tiger II Ausf.A**? To set the record straight, the names found in the original documents have been used in the text and listed in the introductory sections of chapters 4 and 5. As can be seen from these lists, there was no single official name. New names simply evolved with time during the war. However, this does not give us license as postwar historians to fabricate, propagate, use, and spread misnomers. The original German names for the Panzers and component parts are shown throughout the text in bold print. When appropriate, the equivalent wartime American military terms are added in parentheses.

Gruppen-Nummern were used during the war to identify drawings of complete vehicles, sub-components, and parts belonging to each Tiger. As parts were made, they were labeled by either stamping the drawing number into the surface or casting the number onto castings. The numbers were listed in manuals to identify specific replacement parts needed to repair each different **Ausfuehrung** (model). While only a limited application still remains for using these part numbers as an aid in restoring the few surviving Tigers, these numbers are the key to creating accurate models. The detailed **Gruppen-Nummern** lists for each Tiger reveal exactly which component part was adopted from a previous **Ausfuehrung** and which parts were designed specifically for a new **Ausfuehrung** (refer to Appendix C). Because it is very rare to have photographs of all views of a Tiger and the surviving Tigers all have missing parts or postwar changes, these **Gruppen-Nummern** used in combination with the descriptions and drawings of the modifications are essential tools for creating accurate models.

Due to space limitations, the operational characteristics, organization, units, issue records, tactical manuals, operational strength reports, and original experience reports which I normally include with each history of a Panzer have been published in the third volume of this series, entitled Tiger I & II: Combat Tactics.

The author would like to thank Herr Loos and Meyer in the Bunderarchiv/Militärarchiv in Freiburg for their many years of support in providing paths through the massive records. Also Frau Kuhl at the Bildarchiv in Koblenz for her assistance in obtaining photographs of exceptional quality. I was able to gain access to the inside of many of the surviving Tigers with the support and cooperation of the late Colonel Aubry at the Musee Blindes, Saumur, Oberstleutnant a.D. Grundies at the Panzer Museum, Munster, Hauptmann Kosinski at the Wehrtechnischen Studiensammlung, Koblenz; David Fletcher at The Tank Museum, Bovington, and Charles Lemons at the Patton Museum, Fort Knox. This allowed exact identification of each Tiger by its **Fahrgestell Nummer** (chassis number), a significant contribution toward determining the correct sequence of production modifications.

Hilary Louis Doyle was responsible for our efforts at tank diving at the museums. Since 1973, Hilary and I have been partners in digging out details on the design of the German tanks. Hilary is unequaled in the accuracy, detail, and skill applied to the scale drawings illustrating this and many other armor books. Thanks go to Bill Auerbach for reviewing the manuscript and providing photographs. Peter Frandsen, who has accompanied me on many trips to the museum at Aberdeen where new details are always found on the rusting hulks, also reviewed the manuscript, representing the average armor enthusiast. Many other friends including Werner Regenberg, Karlheinz Münch, and Wolfgang Schneider were very helpful in providing photographs.

Once again, special thanks go to my friend and mentor Walter Spielberger, the leader in researching and publishing books on the design of German tanks and other vehicles. In the area of technical descriptions, Walter has led the way. I have merely followed by adding and cleaning up a few details.

Tom Jentz
Germantown, Maryland
April 1997

4

Panzerkampfwagen "Tiger" P2 VK45.02 (P), Porsche Typ 180

The first attempt to design a tank with a long 8.8 cm gun in the turret was undertaken by the team of Porsche and Krupp. Porsche was responsible for designing the chassis and Krupp the turret. Plans were made to mount **8.8 cm Kw.K. L/71** guns in turrets starting with the 101st **Pz.Kpfw.VI (VK 45.01 (P))**. Not bothering with a trial series, contracts for fabrication of the components and assembly of a production series of 100 were awarded already in February 1942. At this time, the first six **VK 45.01 (P2)** with **8.8 cm Kw.K. L/71** guns were scheduled to be assembled and delivered in November 1942. Because of problems with the Porsche-designed engines and suspension, contracts for the production series were canceled on 3 November 1942.

The following listing of official designations is presented as an aid for keeping track of the names as they evolved during this design project:

Porsche designations for their chassis designs:
Panzerwagen-Projekt "Tiger" by 8Oct41
Typ 101 verstaerkt (type 101 strengthened) changed to **Typ 180** by 23Mar42
Typ 181 for hydraulic drive

Wa Pruef 6 official designations:
VK 45.01 (P2) on 4Feb42
Pz.Kpfw.VI (VK 45.01 (P)) (Ausfuehrung P) on 5Mar42
VK 45.02 (P) on 23Mar42
VK 45.02 (P) previously **VK 45.01 verstaerkt** on 24Apr42
Pz.Kpfw.VI, VK 45.01 (P), Tiger (P) by 1Jul42
Pz.Kpfw."Tiger" P2 on 25Jan43

4.1 CHASSIS DEVELOPMENT

The initial concept for the new Panzer was that the chassis designed for the **VK 45.01 (P)** could be adopted with very little modification aside from that needed to fit the new turret and ammunition stowage. But as time went on, views on what was needed on future battlefields caused changes to be initiated in armor protection (with the associated modifications of the hatches, driver's visor, and machinegun ball mount). In addition, Porsche proposed numerous changes to the drive train, including four different engines, two different drives (electric and hydraulic), and both front and rear engine compartments.

On 30 January 1942, all the key players, including Generalmajor Philipps, head of **Wa J Rue (WuG)/VIII** (armaments production department); Professor Porsche, the president of the **Panzerkommission**; Dr. Rohland, the chairman of **Hauptausschuss-Panzer** (a committee of industry leaders to advise on Panzer production); Oberst Fichtner, Oberstleutnant von Wilcke, Oberbaurat Kniepkamp, and Major Crohn of **Wa Pruef 6** (automotive design office of the ordnance department); and Dr. Mueller, Herr Dorn, Herr Woelfert, and Herr Heerlein of Fried.Krupp AG, met to discuss the conceptual design as follows:

*Professor Porsche brought up the request that, starting with **Wanne Nr. 101**, the hulls be fabricated with the front plate at an angle of 60° from vertical. Following discussion of the various proposals and requirements the decision was made that, starting with **Wanne Nr. 101**, Krupp should plan to deliver the hulls with front plates at an angle of 60°.*

[Unfortunately, this document does not include an explanation as to why an angle of 60° was requested by Porsche. The expertise of Dr. Porsche was in designing automotive drive trains, not armor protection. However, as president of the **Panzerkommission**, Dr. Porsche was privy to discussions on changes in all other Panzer design projects, including proposals for the **VK 30.01** series (later named "Panther"). Oberbaurat Rau, the resident expert on armor protection in **Wa Pruef 6**, wasn't present at this meeting. After the war, Rau wrote that there was ample proof in the earliest designs of German armored cars, armored halftracks, and tanks that he knew how to take advantage of sloped plates to increase armor protection - they didn't need lessons from the Russians.]

On 12 February 1942, Daimler-Benz AG, Werk 40, Berlin-Marienfelde wrote to Fried.Krupp:

*In response to our telegram on 11 February you should receive today a new proposal for an entry hatch with built-in driver's visor in a glacis plate set at an angle of 55° from vertical. At the same time we are providing for your information a sketch of a filler piece for a **Kugelblende** 021 B 2782 (machinegun ball mount designed for the **VK 45.01 (H)**). Since we haven't received a dimensional drawing from Porsche, we have assumed that the hatch will be set into a 55o plate because of the external measurements of the large turret, the same as was necessary for a Panzer hull currently being developed by us.*

[Up to this time, the **Pz.Kpfw.III**, **IV**, and **VI** had all been designed with a driver's front plate set at an angle of 9° from the vertical. Therefore, off-the-shelf designs for driver's hatches, driver's visors, and machinegun ball mounts weren't available that could be installed in a plate set at an angle of 55o. Daimler-Benz was awarded contracts by **Wa Pruef 6** to develop these new designs not only for the **VK 45.01 (P2)** but also for their own **VK 30.01 (D)** which was in competition with the **VK 30.02 (M)** designed by **Wa Pruef 6**/M.A.N.]

On 23 March 1942, Prof.Dr.Porsche and Porsche jun. met with Krupp representatives to discuss the **VK 45.02 (P) - Typ 180**:

It was decided that the front plate with a protruding upper rim would be set at an angle of 55°. Until further notice, at least the first 15 hulls would be completed with a driver's visor and machinegun ball mount in the front plate. The joint of the upper and lower front hull plates was to be reinforced by a homogeneous cast armor cap welded in place. The forward section of the belly was to be doubled to 40 mm thick in accordance with drawing 180.53.10.1.
1. A 500 mm diameter hole cut out of the glacis plate was agreed upon for the driver's visor. It was to be located 570 mm from the vehicle midline and 400 mm down from the top edge of the glacis plate.
5. In accordance with drawing number 180.53.10, the upper edge of the upper side plate is to be extended as far above the level of the deck as is possible from the armor plates that have already been rolled.
*9. The borings in the armor plate, to be used for attaching the suspension, idler, and drive sprocket, are exactly the same as the hull for the **VK 45.01 (P)** (**Typ 101** and **102**).*
15. Penetrations for the engine exhaust pipes are not to be made through the hull sides.
*16. Components installed in the fighting compartment are in accordance with the preliminary sketch SK 11145. Components installed in the engine compartment are unchanged from the **VK 45.01 (P)** (**Typ 101**).*
*18. Cuttings into the belly plate are the same as for the **VK 45.01 (P)**.*
19. Openings for filling the fuel tanks are not through the outer walls, but through the deck.
20. The radio antenna is secured with a plate without a socket.
23. Forty-five additional hulls are to be immediately welded together directly following completion of the previously authorized 15 hulls.

The statements in this document give us the first clear picture of the automotive design that Porsche had in mind for his **Typ 180**, designated by **Wa Pruef 6** as the **VK 45.02 (P)**. The drive train was adopted directly from the **VK 45.01 (P)**. Each of the twin Porsche **Typ 101** 10-cylinder, air-cooled gasoline engines, rated at 300 metric horsepower at 2000 rpm, drove an electric generator. The generators were connected in series for more torque and connected in parallel for running at higher speed. Power cables connected the generators to the electric motors mounted at the rear of the vehicle. Each electric motor drove an external wheel with sprockets to propel the tracks. The drive train was designed for driving the Panzer at a maximum speed of 35 kilometers per hour.

On 20 April 1942, the decision was made to increase the thickness of the deck for the hull of the **VK 45.02 (P)** to 40 mm and to add shot-deflecting rails that extended above the front plate and the side plates. [The sides of the turret designed by Krupp were 110 mm above the vehicle's deck. The exact cause for this has not been found in original documents. This may have been needed to provide clearance above the ventilation louvers on the rear deck, to allow space for running the engine exhaust pipes out through the top deck, or to gain additional height above the floor of the hull. Whatever the cause, the problem of protecting the turret ring was to be solved by extending the height of the hull front and side armor plates well above the height of the deck.]

On 8/9 May 1942, during Dr. Porsche's visit to Krupp, the following decisions were made on the **Typ 180**:

The aft upper side wall was to be set at an angle of 15°. The bend at the top is to be relocated from 2400 to 2900 mm so that the entire engine compartment can be enclosed by one large cover.

A sketch was to be prepared for a support for the gun tube during road marches. The support was to be mounted on the rear deflector rail, capable of being slid to two positions - in the center for travel on land and to the side of the exhaust cowling which would be installed for travel under water.

The armor cover for the machinegun ball mount is to be cast.

[This is the earliest record that has been found referring to an armor casting to protect the machinegun ball mount in the sloped glacis plate. As with all changes to their armor designs, **Wa Pruef 6** ordered a full-scale armor replica to be subjected to test firing as proof that the desired level of protection had not been compromised by ports, welds, and fittings. Prior to this, the only practical solution had been to cut a relatively large port into the 55o and 50° sloped glacis plate through which the machinegun could be aimed and fired. When the "letterbox" hinged armor cover was open, its design did not provide protection against a direct hit on the opening or against bullet splash. A spherical-shaped armor casting was later adopted for the Panther, Jagdpanther, Tiger II, and Jagdtiger. The slight loss of protection caused by a penetration of the armor was considered to be offset by the tactical need for secondary armament in the form of a hull-mounted machinegun. The **M.G.34** mounted in the turrets frequently jammed (as reported from 1939 to 1944). It was also reported that the loader was often too busy with other pressing tasks to clear stoppages in the machinegun mounted in the turret.]

On 5 October 1942, the question of how to redesign the driver's visor was resolved by the decision to install periscopes in the **Typ 180**.

4.1.1 HULL DESIGN

The hull design consisted of sloping plates for increased protection. The 80 mm glacis plate was sloped at an angle of 55 degrees, 80 mm front nose plate at 55 degrees, 80 mm superstructure side plates at 15 degrees, 80 mm hull side plates 0 degrees vertical, 80 mm upper tail plate at 60 degrees, 80 mm lower tail plate at 25 degrees, 40 to 25 mm deck plates at 90 degrees horizontal, and 40 mm front belly plate at 90 degrees horizontal and 20 mm rear belly plate at 90 degrees horizontal. The entire rear deck could be removed for maintenance of the engines, generators, and electric motors.

4.1.2 DRIVE TRAIN AND SUSPENSION

The drive train consisted of twin Porsche **Typ 101/3**, 10-cylinder engines connected to electric generators providing power to two electric motors, one for each drive sprocket at the rear of the hull, designed to provide a maximum speed of 35 kilometers per hour. The combat weight of 45 metric tons was distributed over three sets of two 700 mm diameter steel-tired roadwheels per side suspended by lateral torsion bars. The unlubricated 640 mm wide tracks provided an acceptable ground pressure (when the tracks sank to 20 cm) of 1.06 kilograms per centimeter squared.

4.1.3 ADDITIONAL DRIVE TRAIN PROPOSALS

Never satisfied with a single design that could lend itself to rapid mass production, Dr. Porsche pursued new and unproven inventions in his fixation with obtaining the highest automotive performance possible. As shown on Porsche document Sk.7949 dated 5 October 1942, Porsche had initiated conceptual design work for five different versions of the **VK 45.02 (P)** identified as **Typ 180A**, **180B**, **181A**, **181B**, and **181C**. **Typ 180B** was to have a Porsche **Typ 101/4** engine instead of a Porsche **Typ 101/3** engine in the **Typ 180A**. All three **Typ 181** variants were to have Voith II hydraulic drives instead of the electric motors in the **Typ 180** variant. In addition to wider tracks (700 mm instead of 600 mm width), the following engines were to be installed in the **Typ 181** chassis:

- **Typ 181A** with two Porsche **Typ 101/4** ten-cylinder gasoline engines with 15 liter swept volume each generating 300 metric horsepower at 2000 rpm.
- **Typ 181B** with two Porsche Deutz **Typ 180/1** 16-cylinder diesel engines with 19.6 liter swept volume each generating 370 metric horsepower at 2000 rpm.
- **Typ 181C** with one Porsche **Typ 180/2** 16-cylinder diesel engine with 37 liter swept volume generating 700 metric horsepower at 2000 rpm.

Additional data on specifications, dimensions, and capabilities are provided as Tables in Appendix A.

4.2 TURRET DEVELOPMENT

Prior to encountering a single T-34 or KW Russian tank, Hitler demanded that an 8.8 cm gun firing an armor-piercing round capable of penetrating 100 mm thick armor plate at a range of 1500 meters be mounted in a turret, as evidenced from extracts of notes of a meeting held on 26 May 1941:

*In the near term, for the main part we can still employ Panzers with a **5 cm Kw.K.** (tank gun) with success. But we must immediately procure a **"Spitze"** (core) of about 20 heavy Panzers for each Panzer-Division. This amounts to procuring Panzers that:*

- *have increased ability to penetrate enemy tanks,*
- *possess heavier armor than has been previously achieved, and*
- *attain a maximum speed of at least 40 kilometers per hour.*

*The effectiveness of the existing **8.8 cm Kw.K.** and its armor-piercing round were to be increased, so that 100 mm thick armor plate could still be penetrated at a range of about 1500 meters. The 88 had originally been developed strictly for anti-aircraft purposes. Further development of this weapon as a tank gun appeared possible and was to be requested.*

*Design and fabrication of the heavy Panzers, currently under development by Dr.Porsche and the firm of Henschel und Sohn, were both to be accelerated so that, as planned, six of each type will be available for employment in the Summer of 1942. The 8.8 cm gun was still to be mounted in the Porsche-Panzer, but the effectiveness of the gun was to be increased to meet the above specification. There are no objections against mounting the **Rohr 0725** (7.5/5.5 cm tapered bore gun) in the Henschel-Panzer. However, this gun can be produced in quantity only if a satisfactory stockpile of tungsten is available for manufacturing ammunition. Mounting the 8.8 cm gun in the Henschel-Panzer will be investigated.*

On 21 June 1941, the firm of Dr.ing.h.c.F.Porsche K.G. was requested by **Wa Pruef 6** to determine if it was possible to mount the **8.8 cm Flak 41** in the turret for the Porsche **VK 45.01 (P)**. On 10 September, Dr. Porsche reported that only the **8.8 cm Kw.K. L/56** gun could be considered for the **VK 45.01 P**.

After being informed of the status, Dr. Todt, **Reichsminister fuer Bewaffnung und Munition** (head of the ministry for armament and ammunition), sent a letter dated September 23, 1941, to General Ritter von Leeb, head of the **Heeres Waffenamt** (army ordnance department) stating in part:

*The **Waffenamt** has arranged that a turret from Henschel with a turret ring diameter of 1900 mm is to be mounted on the Porsche-Panzer instead of the turret planned by Prof. Porsche with a 2000 mm turret ring diameter. Ever since Prof. Porsche received a design commission from Hitler, I cannot escape the impression that the **Waffenamt** is defending their prestige, so that in the end a **Waffenamt** Panzer (from Henschel) will still be constructed. The **Flak 41** can hardly be installed in a turret with a 1900 mm turret ring. I must inform you that every time I see Hitler, he repeatedly asks if in reality the highly effective **Flak 41** will be installed in the Porsche-Panzer. Prof. Porsche has ensured me that he is striving for possible installation of the **Flak 41**. I cannot relieve him of this task and accept the excuse that the **Flak 41*** *cannot be installed because the **Waffenamt** made arrangements for a narrower turret.*

*Hitler does not feel confident that another 88 mm gun design can be used instead of the **Flak 41**. Hitler wants the **Flak 41** installed in the new heavy Panzer without any degrading modifications. I bring to your attention today that we will have to expect the strongest objections from Hitler if one day during the first demonstration, the Panzer has a gun other than the **Flak 41**.*

Because I am under a directive from Hitler to personally pursue this matter, I request that fundamental changes to the program, such as the elimination of the Porsche turret in favor of the Henschel turret, not be approved without my notification.

The following excerpts from the response of Colonel Fichtner, head of **Wa Pruef 6** (an office in the **Waffenamt** under General Ritter von Leeb) dated September 27 reveal their previous decisions and future plans:

***Wa Pruef 6** has never directed the reduction of the turret ring diameter for the turret for the Porsche-Panzer. The reduction from the original plans by Dr. Porsche for a turret with 2000 mm diameter to the present 1850 mm is solely and exclusively the result of development work by Krupp.*

*In the Spring of 1941, in line with the program to increase weapons capabilities for Panzers, the installation of the **8.8 cm Kw.K. L/56** was planned for the Porsche-Panzer. Dr. Porsche contracted directly with Krupp for the turret for this Panzer, which was developed in close cooperation by both of these firms. In contrast to the usual procedure, the **Waffenamt** did not award a development contract to Krupp.*

In a meeting on July 25, 1941, I informed Prof.Dr.Porsche that I was not satisfied with the Krupp turret and for the future was pursuing an improved solution that would be equally suitable for both the Porsche and Henschel Panzers.

*The firms of Krupp and Rheinmetall are each to receive a contract to present **Wa Pruef 6** with conceptual design projects for a turret fitted with the **8.8 cm Flak 41** for the **Pz.Kpfw. VK 45.01 (Porsche and Henschel)**.*

A new turret was to be designed that would house the higher performance 8.8 cm gun with its longer ammunition, longer recoil, and longer gun needing counterbalancing. [No documentation has been found revealing any attempts by Rheinmetall to create conceptual designs for an **8.8 cm Kw.K.** in a turret for the **VK 45.01** series. Also, records and sketches by Krupp have not been found to fill in the evolutionary sequence from the "horseshoe" shaped turret for the **VK 45.01** series to extending the turret rear and sloping the turret sides to create the design erroneously referred to in postwar literature as the "Porsche turret."]

It appears that final decisions on the shape of the turret had not been made by January 1942, as recorded during the following minutes of an internal meeting within Krupp on 20 January 1942:

In response to the question as to why the turrets developed for the heavy Panzers have a wide front compared to the Russian tanks, Herr Dorn explained that it was completely possible to design a narrow gun mantle but that it would then be relatively taller. That is, the frontal target area taken as the height times the width would be relatively the same for both designs.

[However, comparison of the **VK 45.01** turret with the later designs for the **VK 45.02** and **VK 45.03** shows that further attempts by Krupp to reduce the frontal target area were successful.]

Further meetings held on 15/16 April 1942, attended by **In 6**, **Wa Pruef 6**, Henschel, and Krupp representatives to decide on details reveal that the new turret design was far from solidified:

*Item 4 - It is planned to have an exit hatch on the rear or the side of the turrets for the **VK 45.02 (P)** and **VK 45.02 (H)**. Item 5 - Vision slits are to be available for the gunner and loader in the sides of the turrets for the **VK 45.02 (P)** and **VK 45.02 (H)**. In addition, two forward-facing periscopes (one for the gunner and one for the loader) are to be mounted forward in the roof.*

But the general shape of the turret must have been decided before this time, as armor components were already being fabricated before 7 May 1942 when Krupp reported that they were having problems shaping the **VK 45.01 (P2)** turret fronts: *During heated forging of the turret front plate for the **VK 45.01 (P2)**, 8 out of 15 forged pieces have failed because of cracks on the face at the most stretched locations.* Krupp requested but did not receive permission to repair these pieces for delivery. Instead it was agreed that Krupp would weld the cracks on one piece, reheat it, and then subject the piece to firing trials to determine if the repairs could be made without affecting the armor resistance to penetration. The armor plate for the front plate, made of Krupp alloy PP793, was being shaped to conform to drawing 021 B 48011.

In a meeting on 8/9 May 1942, Dr. Porsche and Krupp decided: *The detailed design of a crane rail or arm on the turret to be used for mounting a lifting rig to aid in removing and installing suspension components awaited data on the lifting strength, length, and traversing arc. The design was also to consider lifting and installing the cover over the engine compartment, which was 2700 mm by 2400 mm and weighed about 1300 kilograms.* [This is the earliest indication of an attempt to install a lifting device on the tank turret to aid in maintenance of the vehicle. This concept didn't get implemented until June 1944 when three **Pilze** (sockets) were welded to the turret roof as the base supports for a **2-ton Behelfskran**.]

Review of the design for the new commander's cupola with periscopes for the **VK 45.02** was completed by Krupp on 21 May 1942. All-round overlapping fields of view were obtained by installing seven periscopes. The most favorable width for the periscopes was determined to be 100 mm. These same periscopes could be installed in the turret roof and on the upper edge of the 30 degree sloped turret sides.

An earlier decision that a turret exhaust fan wasn't needed in a Porsche-Panzer was reversed after problems were experienced with expended propellent fumes after firing the **8.8 cm Kw.K. L/56** gun in the **VK 45.01** turret. On 4 June 1942 a contract was signed for 100 exhaust fans, rated at 12 to 13 cubic meters/minute, for installation in the turret roof of the **VK 45.01 (P2)**.

A letter sent by Porsche to Krupp on 5 June 1942 revealed that the specification for the turret traverse speed when using the electric motor was 15 degrees per second (equal to one revolution in 24 seconds). [This was slower than the contemporary American or British tank turrets, usually rated at one revolution in 15 seconds.]

On 6 July 1942, **Wa Pruef 6** wrote Krupp:

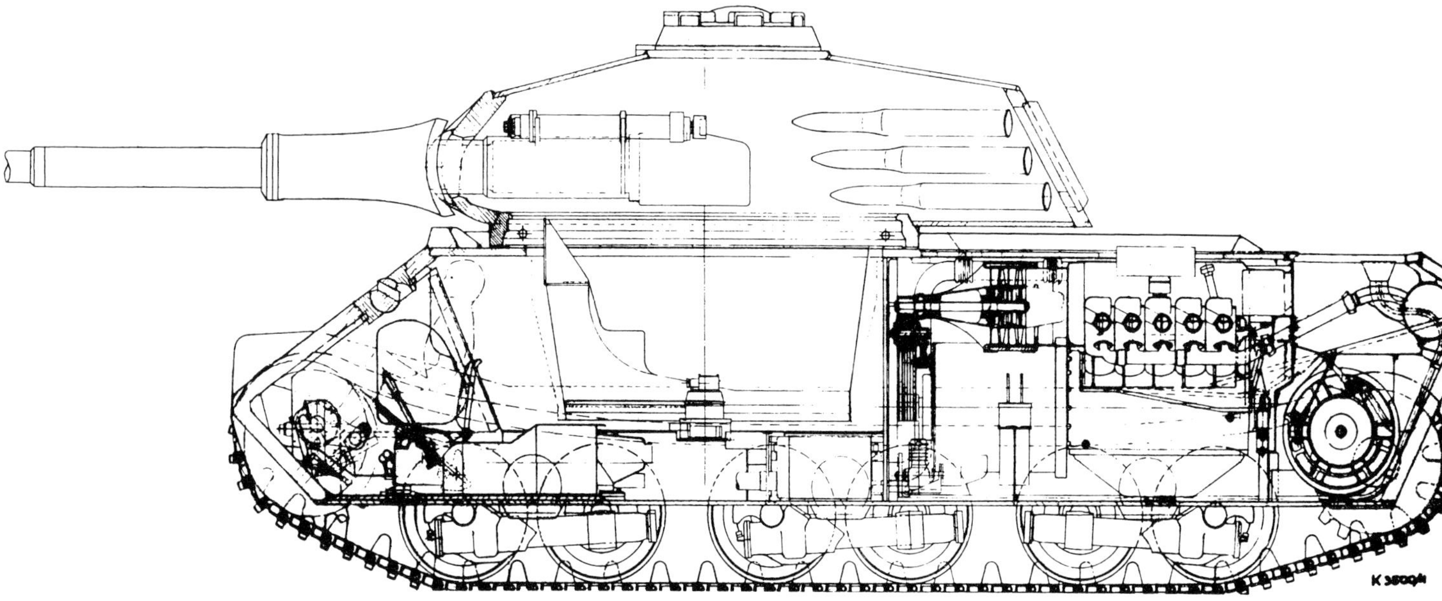

The Porsche-designed **Typ 180** with all drive train components, including the twin V-10 engines and generators, two electric motors, and final drives mounted in the hull rear. Even the exhaust muffler and tail pipe were mounted inside. Unfortunately, details of the braking and steering mechanisms in the forward compartment have faded away in this copy of the original drawing.

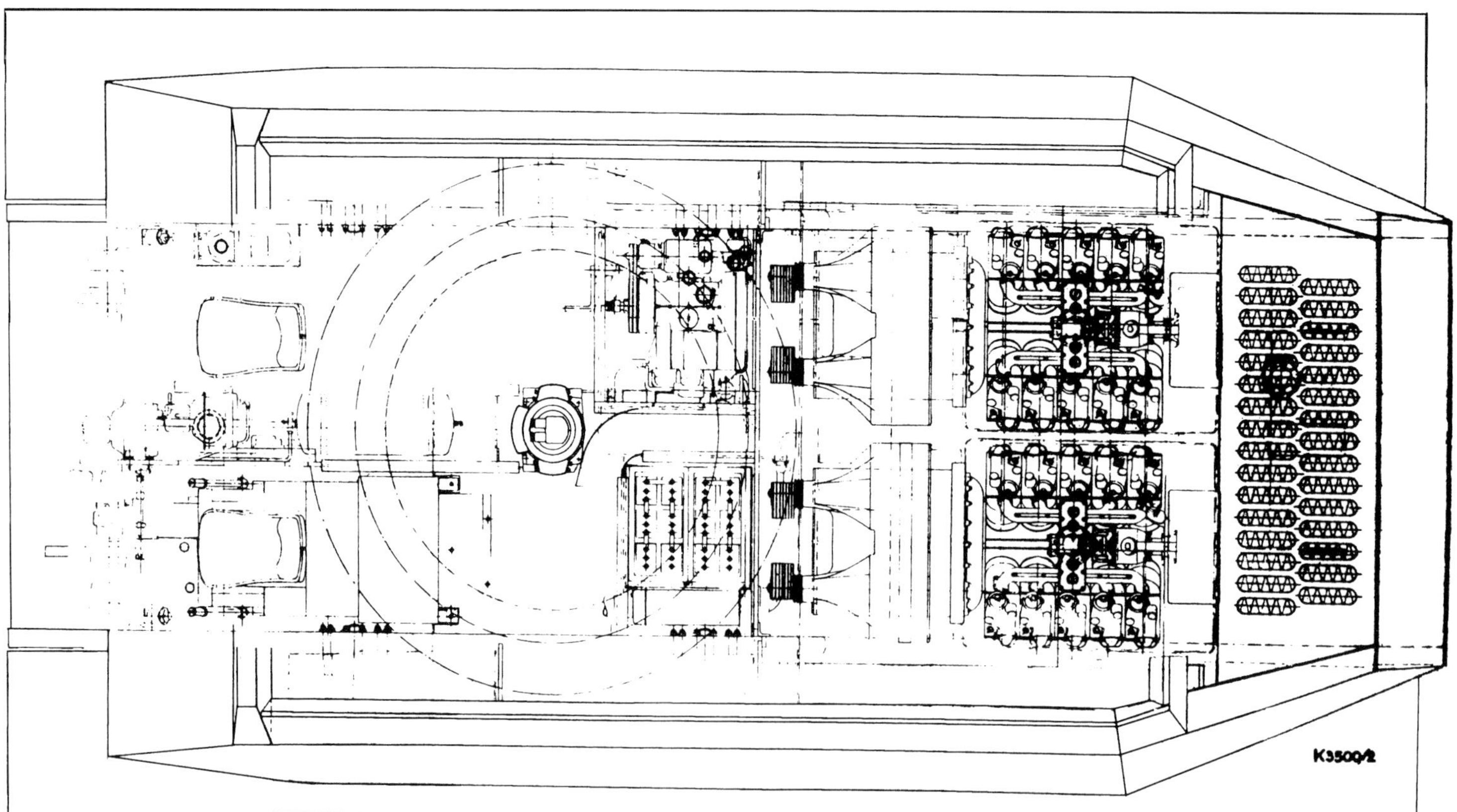

At the conclusion of the demonstration of the commander's cupola with a rotating inner ring for mounting an anti-aircraft machinegun or a scissors periscope, it was decided that:

1. Installation of a rotating ring in the commander's cupola can be eliminated to aid production simplification.

2. The simplest type of mounting is to be designed for the anti-aircraft machinegun. The mount is to be secured on the outer edge of the commander's cupola.

3. In addition, attempts will be made to design a simplified mount for the scissors periscope within the commander's cupola.

***Wa Pruef 6** requested that the rotating ring be dropped from the commander's cupola in newly designed turrets but continue to install the 12 o'clock azimuth indicator ring.*

4.2.1 TURRET DESIGN

The turret design for the **VK 45.02 (P)** that was finally approved by **Wa Pruef 6** for production by Krupp was assigned drawing number sequence 021 St 48011 through 48050. As revealed by the detailed drawing list dated 21 October 1942, every component was newly designed specifically for this turret. Not a single component was adopted from other turrets, not even from the predecessor turrets designed for other heavy tanks.

The curved turret front was designed to reduce the frontal target area to a minimum. A cast gun mantlet attached to the gun sheath covered the opening in the turret front. Two curved armor guards welded to the turret front behind the gun mantlet protected against hits from the side. The turret front was a 100 mm thick plate that was forged into a curved shape with the top at 45 degrees and the bottom at 30 degrees. The sides consisted of two large 80 mm thick plates, which were also bent into curved shapes. With the exception of the thickened bulge (90 mm) under the commander's cupola which was vertical, the turret sides were sloped at an angle of 30 degrees. The removable 80 mm thick turret rear was also sloped at an angle of 30 degrees to the vertical. The turret roof was 25 mm thick at the front at 77 degrees, 40 mm thick in the center at 90 degrees, and 25 mm thick at the rear at 80 degrees. The commander's cupola was an armor casting with slots cut for seven periscopes.

Access to the turret was provided through a hatch in the cupola, a loader's hatch to the right of the cupola, and a hatch in the rear wall. An exhaust fan covered by an armor guard was mounted in the turret roof.

4.2.2 ARMAMENT

The primary armament was the **8.8 cm Kw.K.43 L/71** gun mounted in the gun mantlet. Secondary armament was provided by an **M.G.42** mounted coaxially to the right of the main gun. Machine pistols, pistols, and hand grenades were available for the crew.

4.2.3 AMMUNITION STOWAGE

A total of 68 rounds of ammunition was carried for the main gun. Forty-two rounds were stored horizontally in the panniers along the superstructure sides. Ten rounds were stored horizontally on the floor. Sixteen rounds were stored horizontally in racks at the rear of the turret.

4.2.4 VISION DEVICES AND PISTOL PORTS

Except for the commander, vision for the turret crew was somewhat limited. The gunner had a binocular **Turmzielfernrohr 9b/1** sighting telescope with 2.5X magnification. The loader had a periscope mounted in the turret roof, a pistol port in the right-hand turret wall, and a pistol port in the rear escape hatch. The commander had seven periscopes in the cupola for all-round vision, a pistol port in the left-hand turret wall and a **Schaulock** (view port) to his left.

4.3 PRODUCTION CONTRACTS

On 4 February 1942, **Wa J Rue (WuG 6) VIIId** awarded contract SS-210-5806/41 to Fried.Krupp AG, Essen for 100 **VK 45.01 (P2)** armor hulls and 100 armor shells for turrets. The armor hulls were to be delivered to Nibelungenwerk GmbH, St.Valentin (Niederdonau) for assembly as complete operational chassis under contract SS-210-5911/41. The armor shells for the turrets were to be delivered to Fried.Krupp AG, Essen (Ruhr) for assembling operational turrets under contract SS-210-5910/41. Krupp was to start fabricating and delivering the 100 armor bodies for the **VK 45.01 (P2)** upon completion of the 100 armor bodies for the **VK 45.01 (P1)** under contract SS-210-5803/41. The first eight armor hulls were to be delivered to Nibelungenwerk in August 1942, followed by 15 per month starting in September. The first six operational turrets completed by Krupp were to be delivered to Nibelungenwerk in November 1942, followed by 15 per month starting in December.

Wa J Rue informed Krupp on 5 March 1942 that they were to assemble 100 **Geraet 5 Gr 808** (the drawing number for the **8.8 cm Kw.K.43 L/71**) and DHHV (Dortmund-Hoerder-Hutten-Verein) was to assemble 150. Delivery was to start upon completion of their order for **Geraet 5 Gr 38** (the drawing number for the **8.8 cm Kw.K.36 L/56**) with delivery of the first 15 from Krupp to occur by 1 September 1942.

Following a meeting with Major Kleine of **Wa J Rue (WuG) VIII** on 17 April 1942, Krupp reported:

*After lengthy discussions, **Wa J Rue** representatives stated that they could issue extension contracts for an additional 100 Tiger. A higher number couldn't come into question until after the decision was made if the **Typ P** or the **Typ H** would continue in production. The following contracts were issued verbally: SS 210-5805/42 for 100 **VK 45.01 (P)** armor hulls and armor shells for turrets and SS 210-5909/42 for the assembly of 100 **VK 45.01 (P)** turrets in operational order.*

This contract was confirmed in writing by **Wa J Rue (WuG 6) VIIId** as contract SS 210-5805/41 for a total of 200 **VK 45.01 (P2)** armor hulls and turrets with **Nr. 150101 to 150300**.

During a visit by Dr. Porsche to Krupp on 7/8 May 1942, the following decisions were made on the Typ 180: *Krupp was informed that delivery of the first four **Typ 180** hulls was to be made in October instead of August. To enhance the ability to modify the hulls and installed components, Krupp was requested to initially complete 30 **Typ 180** hulls and only prepare armor plates for the rest.*

The Porsche design for the **Typ 180** with the turret mounted at the rear and the twin V-10 engines mounted in the center.

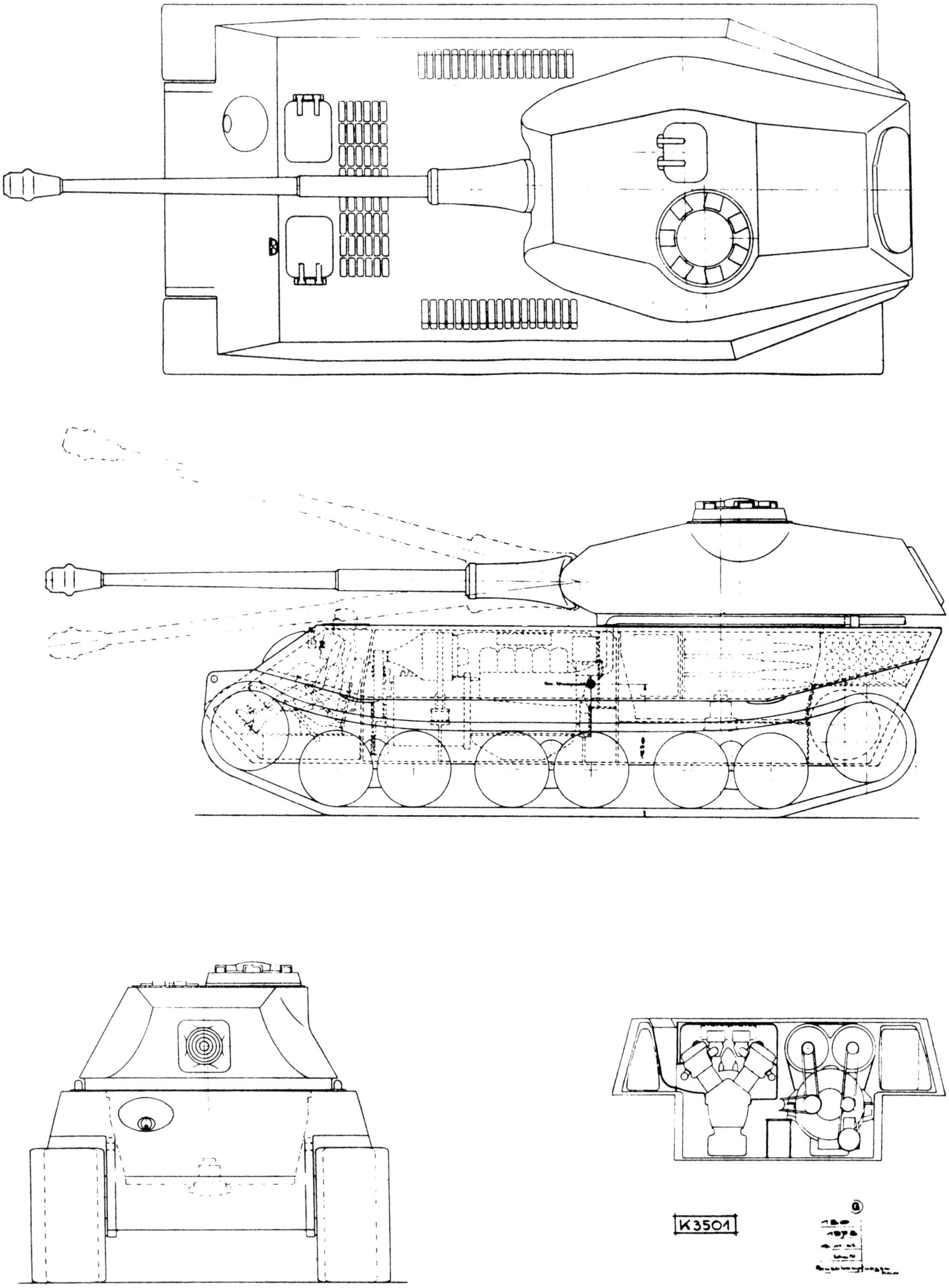

The delivery schedule for operational **VK 45.01 (P2)** from Nibelungenwerk was changed on 13 May 1942 to 6 in December 1942, 15 per month starting in January and running through June 1943, 18 in July, 20 in August, 25 in September, and 30 per month starting in October.

On 21 May 1942, Krupp replied to Porsche's request to delay completion of armor hulls for the **VK 45.02 (P)**, **Typ 180** during the meeting on 8/9 May:

Of decisive importance is your request that at first only 30 ***Typ 180*** *hulls be completed. You have been repeatedly informed that the material for 50 Panzers has already been prepared and for the most part rolled out. On the other hand, you should be aware that we complete 15 each month so that completion of 30 is only significant for two months. Also, if only 30 hulls are completed and the others are to be modified, we assume that these changes will only be minimal in nature and will not affect the outer form of the hulls which were just modified by you. Therefore, we will not interrupt the ongoing work on the 50 hulls already in progress and will complete them in the current agreed upon dimensions.*

By 27 June 1942, 200 **T.Z.F.9d** binocular gun sights for the **VK 45.01 (P2)** had been ordered from Leitz, Wetzlar. [This binocular gun sight was later redesignated **T.Z.F.9b/1** and the **T.Z.F.9d** designation applied to a monocular gun sight.]

On 2 July 1942, Krupp telephoned Nibelungenwerk confirming that the delivery schedule for the **Serie P2** remained unchanged and informing them that completion of the **Probefahrzeug** (trial vehicle), for which Krupp was to deliver the turret in mid-October, would occur on time only if Porsche delivered the required drawings.

Following the failure of the Porsche-designed 10-cylinder engines, on 3 November 1942 **Wa J Rue** informed Krupp that the contracts SS 210-5806/41 dated 4 February 1942 and 18 June 1942 for 100 **VK 45.01 (P2)** armored hulls and turrets were rescinded. A separate contract was to be issued by **Wa Pruef 6** for completion of three **Versuchs-Panzer**.

The question of revival of the production of the Porsche-Tiger was raised as a discussion point for Hitler's conference on 22 November 1942, as follows:

Because production of the first ***Serie Porsche I*** *with the long 8.8 cm gun as a* ***Sturmgeschuetz*** *runs out by 12 May and the* ***Henschel-Tiger*** *with the long gun will not start until the Fall, it should be immediately investigated whether production of about 10 Panzers with 8.8 cm long guns per month can be secured as a further expedient model. The* ***Porsche II Ausfuehrung*** *is a possibility.*

On 27 November 1942, the **Reichsminister fuer Bewaffnung und Munition** stated that in contrast to the previous understanding, only three Panzers of the **VK 45.02 (P)** type were to be completed.

On 25 January 1943, **Wa Pruef 6/IId** awarded Krupp, Essen contract SS 006-4599/42 for three operational turrets for **Pz.Kpfw."Tiger" P2**, including assembly with **8.8 cm Kw.K.43** guns, and contract SS 006-6399/42 for three **Pz.Kpfw."Tiger" P2** armor hulls to be shipped to Nibelungenwerk. On 28 January, Krupp informed **Wa Pruef 6** that the three hulls for the **Versuchsfahrzeuge Tiger P2** had already been delivered to Nibelungenwerk and that the turret armor bodies had already been completed.

On 17 February 1943 Krupp reported that, based on information provided by Prof.Dr. Porsche:

Three ***Tiger 2 (VK 45.02 (P))*** *with electrical drive and new suspensions are being completed at Nibelungenwerk. The production series is to have hydraulic drives, a new suspension, an air-cooled diesel engine rated at 900 metric horsepower, increased armor thickness, and heavier armament. Delivery is scheduled far in the future.*

On 25 April 1944, Krupp reported on the status of assembly of the three Tiger P2 turrets. One turret had been completed. While all components for both of the other turrets had been completed, the accompanying tools and equipment weren't available. Krupp promised maximum effort to complete these three turrets for use in the Tiger H2-Serie. Krupp reported on 22 August 1944 that at the request of **Wa Pruef 6**, all three **VK 45.01 (P2)** turrets had been converted to the Henschel model with hydraulic traverse.

5

Tiger II Development

Henschel was the third company involved in developing a heavy tank mounting the long 8.8 cm gun. Their entry into the race came later than Porsche. A decision had been made by 5 March 1942 that following the first 100 **VK 45.01 (H)** with **8.8 cm Kw.K. L/56** guns, Henschel's second series were to have **7.5 cm Kw.K. L/70** guns mounted in the turret. It was not until April 1942 that Henschel started conceptual design work on a chassis for a turret with the **8.8 cm Kw.K. L/71** gun. Their first attempt, designated as the **VK 45.02 (H)**, was a makeshift design which didn't survive very long even on the drawing boards. Design work on their second attempt, designated as the **VK 45.03**, started in earnest in October 1942. At first the design of the **VK 45.03** was still based on many automotive components borrowed from the **VK 45.01 (H)**. But in February 1943 Henschel was ordered to redesign the **VK 45.03** in order to use as many standardized parts as possible that could be shared with the M.A.N. design for the Panther II.

Krupp was the sole designer of the turrets for the **VK 45.02 (H)** and **VK 45.03**. The only difference between the turret designed by Krupp for the **VK 45.02 (P)** and the **VK 45.02 (H)** was that turrets mounted on Porsche chassis had electric traverse drives while turrets mounted on Henschel chassis had hydraulic traverse drives. The 50 turrets originally fabricated by Krupp for the **VK 45.02 (P)** were modified, fitted with hydraulic drives, and mounted on the first 50 **VK 45.03** chassis from Henschel. Krupp also designed the "**Serien-Turm**" that was mounted on all subsequent Henschel chassis starting with the 51st **VK 45.03 (Fgst.Nr.280048)**.

The following listing of official designations is presented as an aid for keeping track of the names as they evolved during this design project:

<u>**Wa Pruef 6** designations:</u>
VK 45.02 (H) by 16Apr42
Tiger II for the **VK 45.02 (H)** by 18Sep42
Tiger III (VK 45.03) by 12Oct42
Henschel Tiger B on 8Jan43
Tiger II for the **VK 45.03** by 3Mar43
Pz.Kpfw.Tiger Ausf.B on 2Jun43
Pz.Bef.Wg.Tiger Ausf.B on 2Jun43

<u>**In 6** official designations:</u>
Pz.Kpfw.Tiger (8,8 cm Kw.K. L/71) (Sd.Kfz.182) by 5Mar43

<u>**Reichsministerium fuer Bewaffnung und Munition** names:</u>
Koenigstiger by 11Dec44

<u>**Henschel** designations:</u>
VK 45.03 by 28Oct42
Tiger III by 10Dec42
Tiger 3 by 2Feb43
Tiger II on 5Mar43
Tiger 2 by 8Mar43

5.1 CHASSIS DEVELOPMENT

The initial concept for the new **VK 45.02 (H)** with a **8.8 cm Kw.K. L/71** gun in the turret was that many components designed for the **VK 45.01 (H)** chassis could be adopted with very little modification. The drive train and suspension components were to be retained, but the hull was to be redesigned with sloped armor protection. At the time there wasn't any machinegun ball mount designed for a glacis plate sloped at 50 to 60 degrees from the vertical. Therefore, during a meeting in Kassel on 15/16 April 1942 attended by representatives from **In 6**, **Wa Pruef 6**, Henschel, and Krupp, it was decided that:

*A built-in mounting for the hull machinegun is not envisioned for the **VK 45.02 (H)**. Instead, there is to be a port through which a machinegun can be shoved and rested for firing.*

[This was a parallel design to the machinegun port in the glacis of the **Panther Ausf.D** and **Panther Ausf.A** produced up to November/December 1943.]

On 19 August 1942, a requirement to install the Panther engine (Maybach HL 230 P30) into the Tiger caused Maybach to suggest an interim design. Starting with production series chassis number 201, the **VK 45.02 (H)** could be outfitted with the Panther engine and Panther cooling system. According to Maybach it would only be necessary to slope the hull rear at an angle of 22

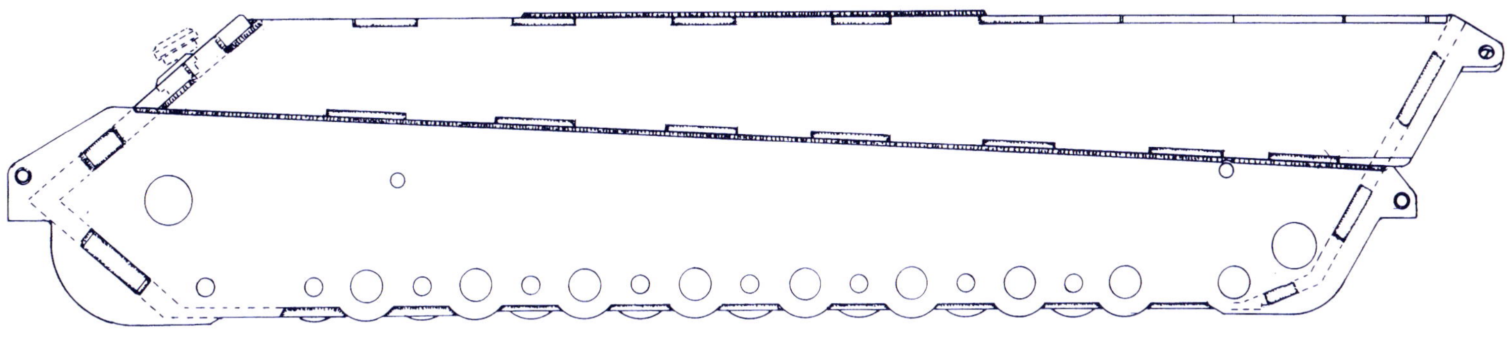

ABOVE: As shown in HSK Nr.J3104 dated 25Nov42, the original glacis plate for the **VK 45.03** was only 100 mm thick. HSK Nr.J3115 dated 30Nov42 for the **VK 45.03** still had the same style of driver's view port with armor visor as the **Panther Ausf.D**

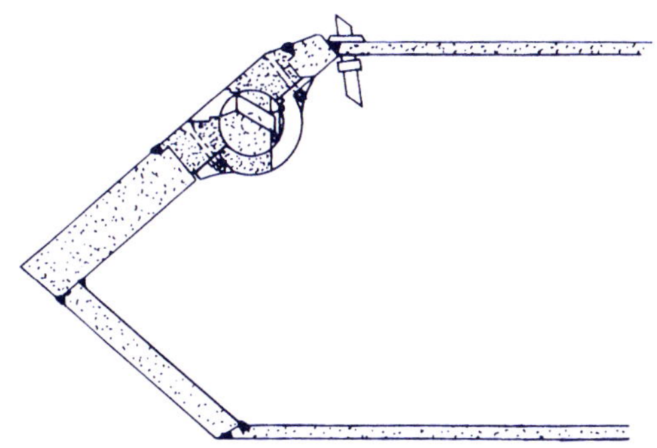

RIGHT: A **Fahrersehklappe (Walze)** (rotating cylinder with direct vision slit) for the **VK 45.03** was proposed by Henschel in HSK Nr.J3174 dated 4 January 1943.

degrees from the vertical [necessitated by the different design of the fan drives between the Tiger and Panther engines]. Henschel started work on the conceptual design.

At the end of August 1942, Reichsminister Speer decided to extend Henschel's contract for assembling production series Tiger chassis by an additional 300 to a total of 600. The first 140 of the additional 300 were to be fabricated exactly the same as the current model. The last 160 Tigers were to retain the same drive train and suspension as the current model, but they were to have sloped frontal armor. Maybach's proposal to introduce an interim type with a Panther rear, engine, fan drives, and cooling system was rejected.

Proof that Henschel was working on a chassis for the **Tiger H3** as early as August 1942 is revealed in the record of a meeting between **Wa Pruef 6** and Krupp representatives on 19 August 1942 concerning the **Tiger-Turm (H3 und P2)**:

Henschel requested that the design of the turret platform supports be examined to determine if they could be moved in about 10 to 20 mm because this would allow a significant increase in the amount of ammunition that could be stowed in the hull side panniers.

In early October 1942, plans for production of the **VK 45.03** were reviewed. It was considered intolerable that such a large number as 424 **VK 45.01 (H)** needed to be produced before Henschel could start **VK 45.03** production. Henschel recommended that they produce 330 **VK 45.01 (H)** and then switch production to 170 **VK 45.02 (H)**, so that **VK 45.03** production would start at number 501. Oberst Thomale (with the **Panzerkommission**) rejected this plan to insert a **VK 45.02 (H)** between the **VK 45.01 (H)** and **VK 45.03**. He consented to starting **VK 45.03** series production in September 1943 and agreed to this providing 100 Tiger H3 for the Spring offensive in 1944.

In a meeting on 12 October 1942, Oberbaurat Kniepkamp related:

The Tiger II is not going to be produced. Start of the Tiger III (VK 45.03) is requested for July 1943 but is probably not possible until September 1943. The Panther engine and cooling system is to be installed. Interchangeable installation of a Maybach-Olvar, ZF-Alklauen, or electro-magnetic transmission; in addition a fully automatic AEG three-speed hydraulic transmission. Suspension with doubled instead of tripled interleaving of the roadwheels. Addition of another roadwheel and use of 1 meter wide tracks to maintain the ground pressure at 0.7 kilograms per centimeter squared. Interchangeable suspension with rubber-cushioned steel-tired roadwheels. The current roadwheels with rubber tires on the Henschel-Tiger only allow a continuous maximum speed of 25 kilometers per hour on the Autobahn.

Henschel completed sketch HSK Nr. J3018 on 19 October 1942 showing the roadwheel layout for their **VK 45.03**. Four rubber-tired roadwheels per axle running on a 760 mm wide track was proposed. Every other axle had two outer pairs of roadwheels and two inner pairs; the interim axles had four centered roadwheels together. The roadwheel design was the same basic design as for the **VK 45.01 (H)** with rubber tires having a 70 mm base width and 800 mm outer diameter. It wasn't until 17 January 1943 that Henschel completed sketch HSK Nr. J3198 showing their proposal for steel-tired rubber-cushioned roadwheels for the **VK 45.03**.

As a result of discussions in Berlin on 25 November 1942, Henschel prepared a new sketch HSK Nr. J3104 dated 25 November 1942 showing the front and rear armor protection for the **VK 45.03**. The 100 mm thick glacis plate was angled at 50 degrees, 100 mm lower hull front plate at 50 degrees, 80 mm superstructure side plates at 30 degrees, 25 mm pannier floor plate at 90 degrees, 80 mm hull side plates at 0 degrees, 80 mm hull rear plate at 30 degrees, 60 mm lower hull rear plate at 68 degrees, 40 mm forward belly plate at 90 degrees and 25 mm rear belly plate at 90 degrees.

Henschel's sketch HSK Nr. J3115 dated 30 November 1942 shows the proposed front of the **VK 45.03** with the same style of driver's view port with armor visor cut into the glacis as in the **Panther Ausf.D**. A large rectangular (760 mm wide by 1700 mm long) hatch was cut into the hull roof to allow access for removal of the transmission and steering unit without removing the turret. Two rectangular (360 mm wide by 500 mm long) hatches were cut into the large maintenance hatch cover plate, one for the driver and one for the radio operator.

Drawings were also prepared by Henschel showing the proposed installations of the **Lenkgetriebe L801** (Henschel L801 steering unit) in the **VK 45.03** on 8 December 1942, the 8-speed Maybach **OG 40 16 36** transmission fitted on 24 October, the 10-speed electro-magnetic transmission fitted on 28 October, and 7-speed Zahnradfabrik **AK 7-200** transmission fitted on 26 November.

In Speer's conference with Hitler on 3 January 1943 as discussion item 7 it was recorded that: *For the Tiger, Hitler has decided that from the start the new Tiger, currently being designed with the long 8.8, is to have 150 mm thick armor on the front and 80 mm thick armor on the sides.* This decision was implemented in a sketch from Henschel HSK Nr. J3168 dated 1 Jan 1943 (effective 9 January 1943) of the **VK 45.03 Bug mit verstaerkter Stirnwand** (front with strengthened front plate). The sketch shows the glacis plate increased to a thickness of 150 mm at an angle of 50 degrees to the vertical and the protective ring segment for the final drives increased in thickness from 80 to 100 mm. The front lower hull plate remained at 100 mm thick at an angle of 50 degrees. Increasing the thickness of the glacis and final drive guard caused a 1760 kilogram increase in the weight.

Henschel was still attempting to solve the problem of vision devices for the driver that would not degrade the armor protection. In sketch HSK Nr. J3174 dated 4 January 1943 Henschel proposed a **Fahrersehklappe (Walze)** that was a rotating cylin-

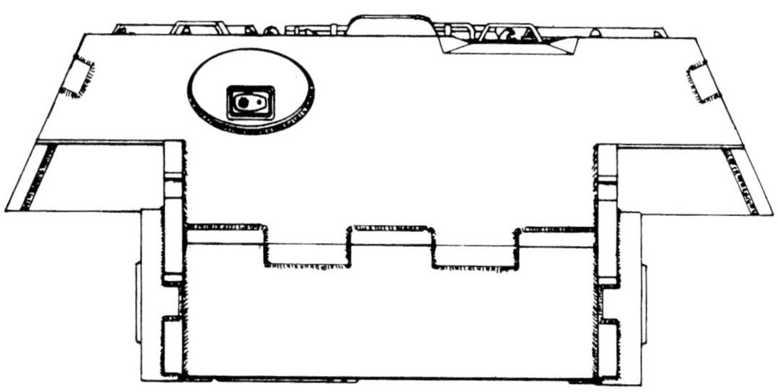

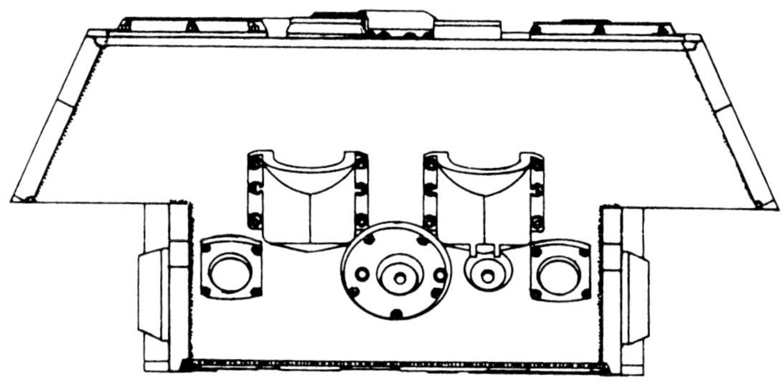

THIS PAGE AND OPPOSITE TOP: Drawing Nr.021 B 49501 dated 12Jan43 shows the armor layout as accepted for production. The 150 mm thick glacis plate was set at an angle of 50 degrees from vertical. The hull rear, made from a single 80 mm thick plate, was set at an angle of 30 degrees.

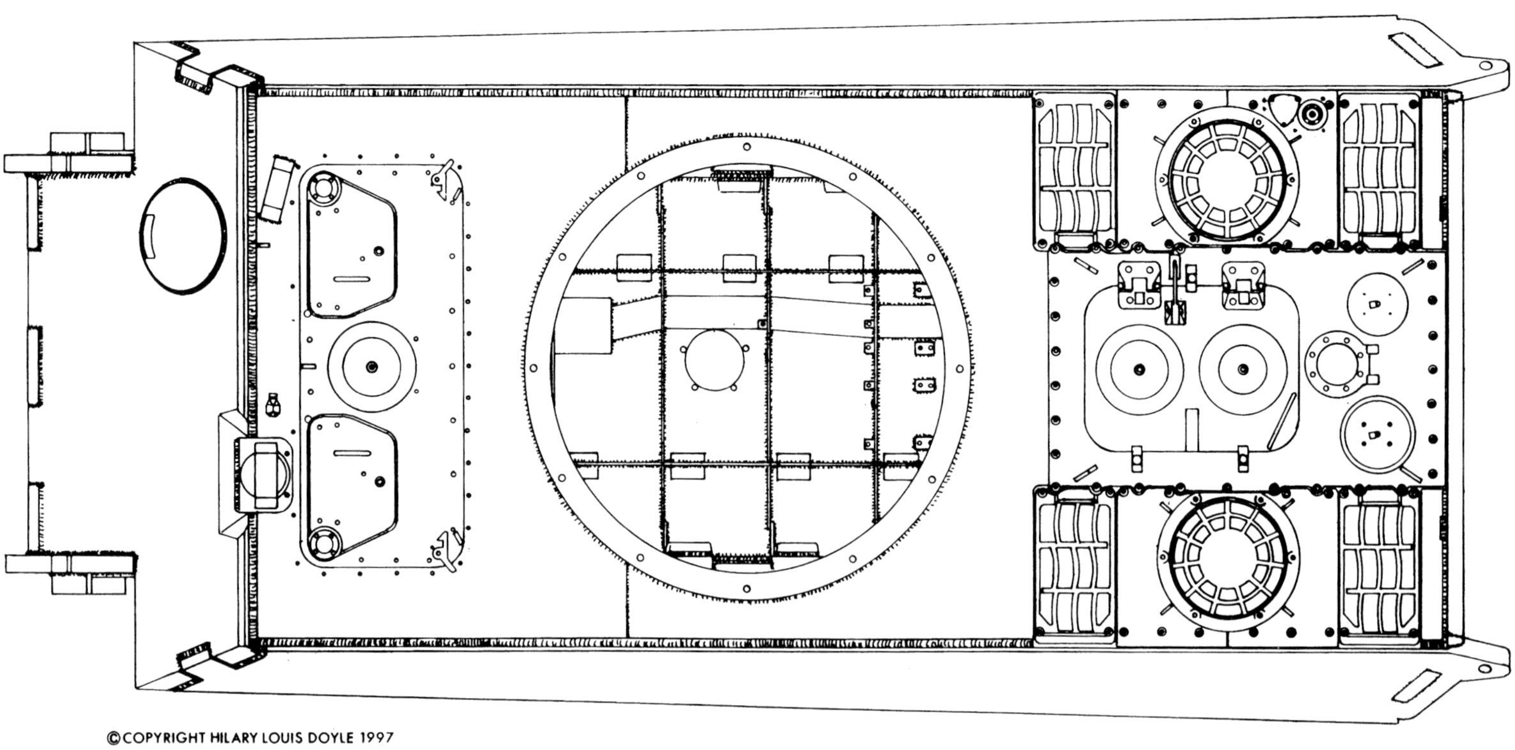

der with direct vision slit. Sketch HSK Nr. J3176 also dated 4 January 1943, proposed a periscope in addition to a **Fahrersehklappe** that was similar in design to the driver's vision port on the Panther Ausf.D. It wasn't until 12 January 1943 that drawing 021 B 49501 was created showing the front of the **VK 45.03** hull as it was later assembled with the **Kugelblende fuer MG** (machinegun ball mount) in its cast armor cover and a **Winkelspiegel** (periscope) for the driver without a visor cut into the glacis plate.

On 17 February 1943, during a meeting at Speer's **Reichsministerium**, it was agreed that the **Tiger 3** (later redesignated the **Tiger II**) should be standardized with the **Panther 2**. Porsche believed that the **Einradien-Lenkgetriebe** (single-radius steering gear) would be satisfactory. The **Zweiradien-Lenkgetriebe** (double-radius steering gear) was to be installed if reports from Direktor Blaicher confirmed that it was ready for mass production. Both the Tiger and Panther were to receive the **Zahnradfabrik Friedrichshafen AK 7/200** transmission, the **Maybach HL 230 Motor** with cooling system and steel-tired, rubber-cushioned roadwheels. The **Panther 2** was to have seven roadwheels and the **Tiger 3** nine roadwheels, but each with different suspension arms. The 660 mm wide combat tracks for the **Panther 2** were to be utilized as transport tracks for the **Tiger 3**.

In his **Anmerkungen zu Tiger B** written in February 1945, Dr.Ing. Aders, head of Henschel's design office, noted:

*In February 1943, **Wa Pruef 6/III** required thorough compatibility between the Tiger B and the Panther II that was also in development by M.A.N., Nuernberg. The highest possible number of complete components were to be exchangeable. Meetings were held and a designer from Henschel worked for a while at M.A.N. What was achieved was that the following components designed for the Tiger B were taken over by the Panther II: L801 steering unit, the final drives, and the Argus brakes. The Tiger B had to take over from the Panther II the engine cooling system, engine compartment, transmission ventilation, fuel system, ventilation for the engine exhaust pipes, engine compartment deck, engine exhaust system, and turret hydraulic drive. These changes caused the completion of the design to be delayed by about three months.*

In sketch HSK Nr. J3315 dated 23 March 1943, Henschel proposed an **Abweissung um Turm bei "Tiger H4"** (deflector around the turret) to protect the exposed turret ring of the first 50 turrets. The segmented armor ring (80 mm base and 60 mm high) with a 1200 mm radius was to be bolted to the hull roof.

In sketch HSK Nr. J3319 dated 24 March 1943, Henschel proposed the motor compartment for the **VK 45.02** with the rear deck layout exactly as it was fabricated for the production series. The hull rear plate was shown to extend all the way down to the belly plate, the 60 mm lower hull rear plate sloped at 68 degrees having finally been eliminated. After the Tiger H3 was renamed the Tiger II, Henschel had made the mistake of correspondingly changing the VK designation to **VK 45.02** on some of their proposal sketches during the period from 23 March to 9 May 1943. On 8 April 1943, Henschel's sketch HSK Nr. 3340 for a proposed adjustable driver's seat was again identified with the correct designation as being for the **VK 45.03**. The driver's seat in the Tiger II

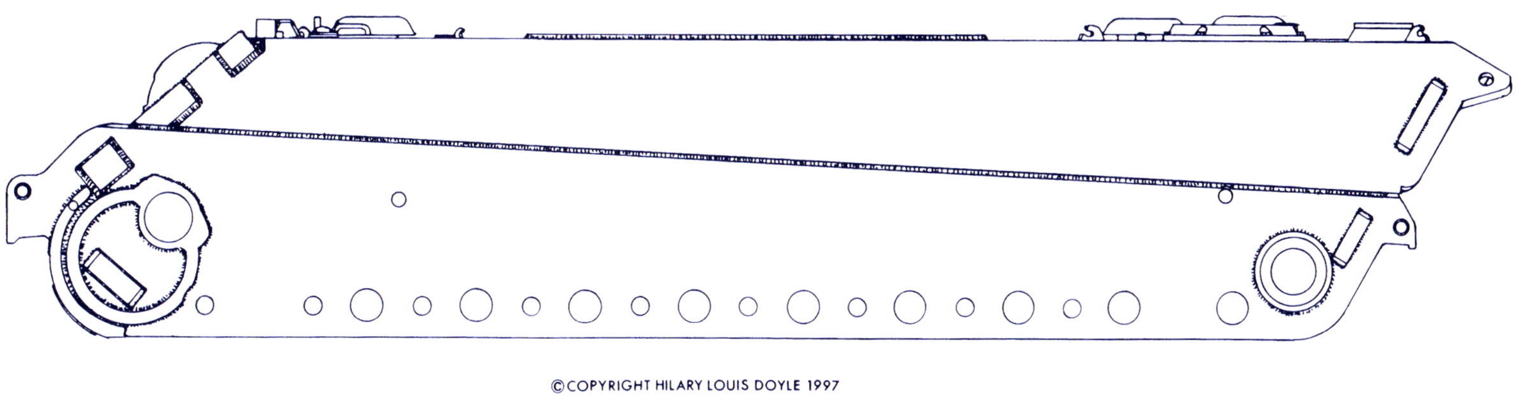

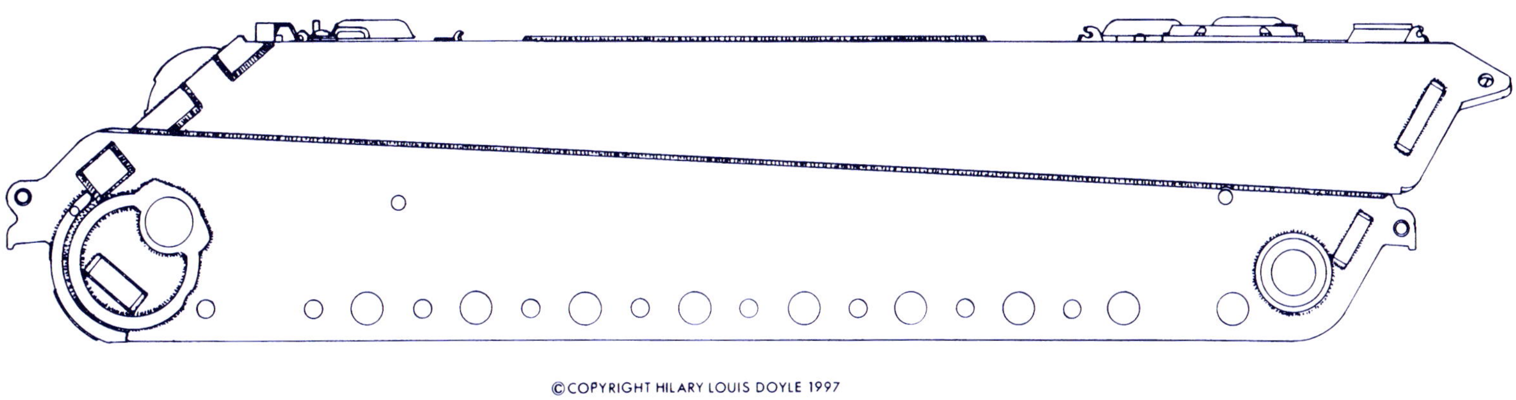

ABOVE: Starting in April 1944, the face of the hull side extensions at the front and rear, drilled for mounting tow shackles, were cut out to allow additional room for fitting C-hooks.

could be raised so that the driver's head protruded above the hatch or lowered to a position where the driver could use his periscope when buttoned up.

In a memoir entitled **Die Entstehung der Fahrzuege Tiger B** dated 6 February 1945, Dr.Ing. Aders, head of Henschel's design office, stated:

The following components were newly created by Henschel for the Tiger B: hull with engine compartment deck, new crew hatches, steering unit, final drives, suspension with drive sprocket wheel and idler wheel, ammunition racks, stowage for parts and equipment, adjustable height driver's seat, and driver's periscope that could be turned sideways and elevated.

The following components were subcontracted or developed by others: HL 230 P30 engine and OG 40 12 16 B transmission by Maybach-Motorenbau, tracks by Ritscher-Moorburg, brakes by Suedd.Argus-Werke, rubber-cushioned steel-tired roadwheels by Deutsche Eisen-Werke, turret with gun by Fried.Krupp, and the machinegun ball mount by Daimler-Benz AG.

The chassis design for the **VK 45.03** that was finally approved by **Wa Pruef 6** for production by Krupp was assigned drawing number sequence 021 St 49501 through 49599. As revealed by the detailed drawing and parts list, practically every component was newly designed or modified for this chassis. Even though Dipl.Ing. Jaeger in **Wa Pruef 6** had specified that only tested and proven component designs were to be used in the **VK 45.03**, this was rarely the actual case. The only component taken over from its predecessor the **VK 45.01 (H)** was the OG 40 12 16 transmission, which was modified by removing the power takeoff for the turret hydraulic drive. Very few parts were taken over unmodified from other Panzer designs. Exceptions were the 800 mm diameter rubber-cushioned steel-tired roadwheels (021 St 50204) and the radiator filler tank (021 St 50235) from the Panther II, and the water and fuel filters (021 St 51152) from the Panther Ausf.G.

Refer to Chapter 6.1 for further detailed descriptions of the **VK 45.03** chassis. Additional data on specifications, dimensions, and capabilities are provided as Tables in Appendix B.

5.2 TURRET DEVELOPMENT

Development of a turret for the **VK 45.02 (H)** ran parallel with the turret for the **VK 45.02 (P)**. As originally conceived, for all practical purposes these were the same turrets, differing only in that the turret drive for the **VK 45.02 (P)** was to be an electric motor and for the **VK 45.02 (H)** an hydraulic motor.

Meetings held on 15/16 April 1942, attended by **In 6**, **Wa Pruef 6**, Henschel, and Krupp representatives to decide on details, reveal that the two new turrets were designed in parallel as follows:

*Item 4 - It is planned to have an exit hatch on the rear or the side of the turrets for the **VK 45.02 (P)** and **VK 45.02 (H)**. Item 5 - Vision slits are to be available for the gunner and loader in the sides of the turrets for the **VK 45.02 (P)** and **VK 45.02 (H)**. In addition, two forward-facing periscopes (one for the gunner and one for the loader) are to be mounted forward in the roof.*

Krupp's annual report for business year 1941/42 (October 1941 to September 1942) stated that the following guns and turrets were still in the design process:

- **8.8 cm Kw.K. L/71** with turret with electrical traverse for the **Porsche-Wagen (Tiger P2)**
- **8.8 cm Kw.K. L/71** with turret with hydraulic traverse for the **Henschel-Wagen (Tiger H3)**

In a meeting on 19 August 1942 concerning the **Tiger-Turm (H3 und P2)**, **Wa Pruef 6** and Krupp representatives discussed redesigning the turret to simplify construction:

Examination of the turret drawings revealed that the design was in no way suitable for mass production. The armor shell for the turret alone required a very large expenditure of man-hours. The following points must be undertaken in a thorough redesign to reduce fabrication time:

a. Interlocking the turret side walls with the turret front must occur so that as little fitting work as possible is needed. Fitting and machining of three 80 mm wide surfaces with a hand-held grinding machine is considered to be intolerable.

b. Machining the plates on the other contact points must be reduced to as little as possible.

c. The shape of the turret rear with its numerous surfaces, which need to be machined for the hatch to dismount the gun, must be completely redesigned.

d. The gun mantle must be redesigned with the goal of reducing complicated boring and machining.

After cancellation of the contracts for the Tiger P2 in early November 1942, decisions had to be made on utilization of parts already fabricated for these turrets. At a meeting between Krupp and Wegmann representatives on 8 December 1942 in regard to assembly of Tiger H3 turrets it was decided that:

At Fried.Krupp a series of components are being worked on for 100 Tiger P2 turrets (including the four Tiger P2 turrets that are to be completed) that can be taken over without modification for installation into the Tiger H3 turrets. Because fabrication is very far advanced, it was agreed that these components be completed by Fried.Krupp and then taken over by Wegmann. The designs still need to be reviewed to ensure that these components match the current designs for the Tiger H3 components. The other components (including the elevation gear) won't be completed by Fried.Krupp.

Further developments in the design of the Tiger H3 turret were discussed in a meeting between **Wa Pruef 6** and Krupp on 10 December 1942:

*The requirement to move the **Schildzapfen** (gun trunnions) farther forward, in order for the apertures for the gun sight and the machinegun to be as small as possible, is contradictory to the requirement for the simplest forging method for the turret front plate. The simplest forging method has priority. One hundred turret front plates with the sharply curved form are in construction. The form of the turret front, starting with the 101st turret, hasn't been decided on because it is feared that it will interfere with the driver's and radio operator's ability to get out.*

*The first model of the **"Russen-Kegels"** (Russian ball) for a pistol port meets the requirement for ease of operation. Its resistance to armor-penetrating hits hasn't been tested yet.*

The gun travel lock at an elevation of 14 degrees is to be completed in accordance with Krupp's proposal AKF 31575. The

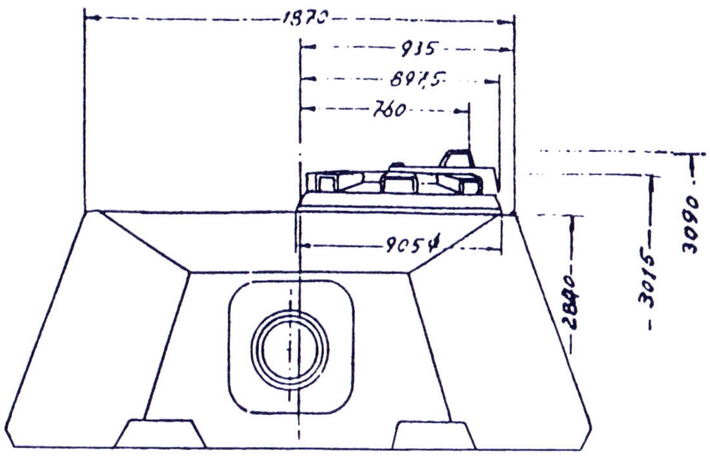

Preliminary views of the new simplified turret design without a side bulge for the commander's cupola and with a sloped front plate as shown in Krupp drawing 2 AKF 31702, incorporated into Henschel drawing HSK Nr.3470 dated 3 June 1943.

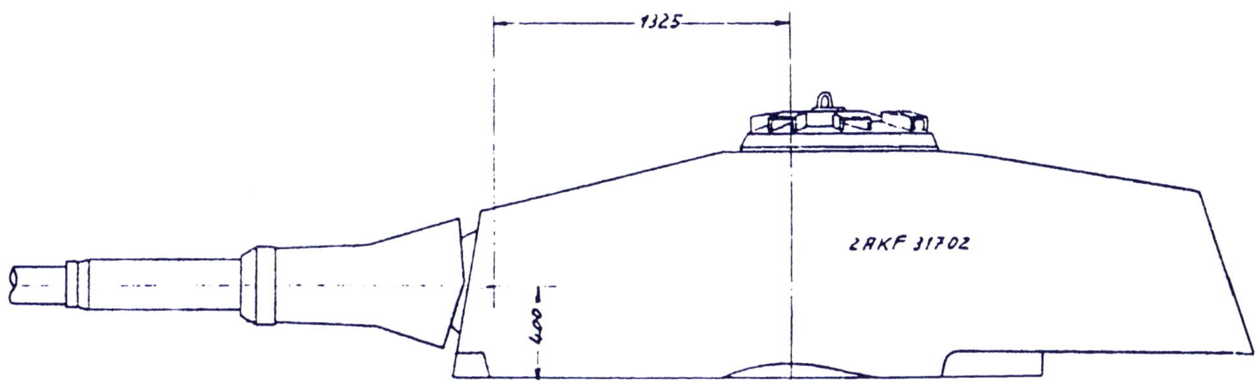

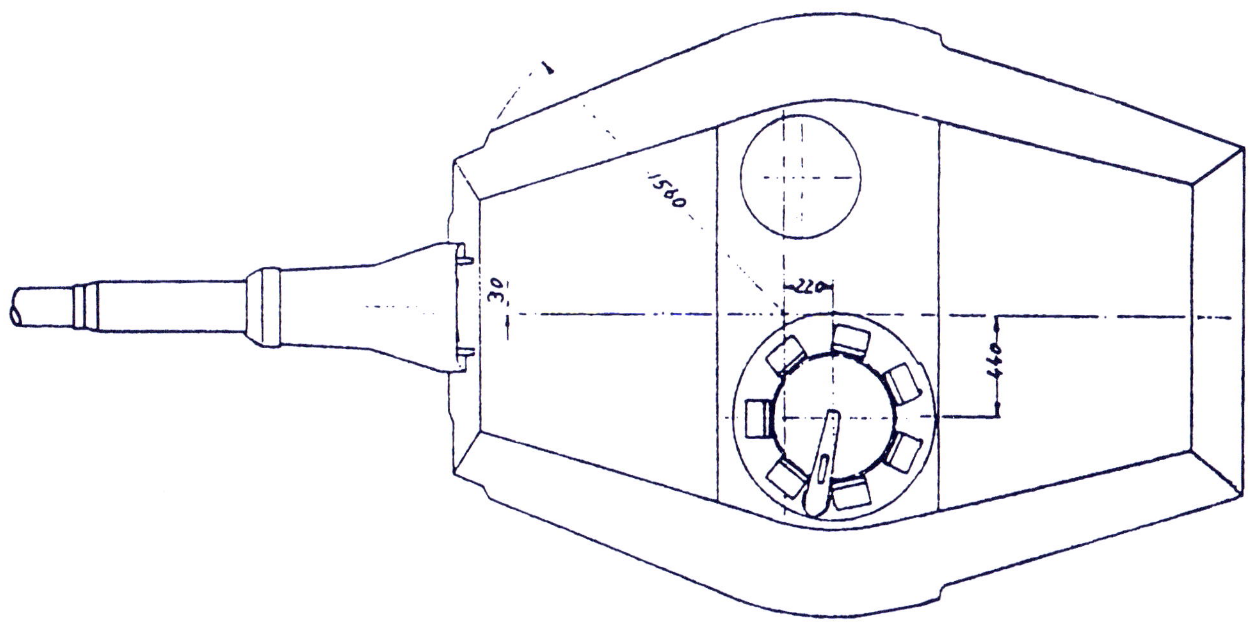

ability to seal the gun mantlet for underwater travel must be ensured when the gun is held at 14 degrees elevation.

A ventilator fan rated at 12 m^3/min is to be built into the rear turret roof in order to clear out the smoke in the fighting compartment. Instructions on installing the ***Ausblasevorrichtung*** *(bore evacuator) are to follow.*

The selection of turret traverse speeds was discussed by representatives of **Wa Pruef 6/IId**, Krupp and Henschel on 15 December 1942:

At the time of the development of the Tiger H1, the traverse speed for the electrical motor driven Tiger P1 turret was specified as 8°/second. Krupp then attempted to meet this specification in the hydraulic motor traversed Tiger H1 turret whose speed was variable based on the speed of the vehicle's engine. Krupp chose a medium engine speed of 1750 rpm for achieving the turret traverse speed of 8°/second. Turret traverse speed of 12°/second could be achieved at the engine's top speed of 3000 rpm. The requirement to traverse the turret at 6°/second when the vehicle's engine is idling at 600 rpm didn't appear until later.

Further design changes to the Tiger H3 turret were discussed by Krupp and **Wa Pruef 6** representatives in a meeting on 15 January 1943 as follows:

Item 1 - The ***1. Versuchsturm*** *with gun and commander's cupola with periscopes is to be sent by 1 February to* ***Wa Pruef 6*** *in order for them to examine the interior layout and operability of the turret.*

Item 2 - The ***2. H3 Versuchsturm*** *is to be delivered for use as a target for firing trials - complete with all components installed with the exception of a shortened gun tube instead of a gun.*

Item 6 - Reinforcing the gun mantlet in pursuing the further development toward 150 mm rolled or forged material results in an estimated weight increase of 300 kilograms. Reinforcing to about 180 mm, corresponding to an 150 mm thick turret front plate sloped at 50 degrees, results in a weight increase of 500 kilograms. In this design the gun trunnion mounts must be moved closer to the gun, resulting in the elimination of the bulge needed for the previous forward-lying gun trunnion mounts. Krupp is to clarify with DHHV the shape in which this new turret front plate can be manufactured.

Item 7 - The investigation with a 100 mm thick turret front plate at an angle of 20° from the vertical revealed that the driver's and loader's hatches could not be opened with the turret positioned from 10 to 2 o'clock. Even by raising the turret lower edge to 110 mm above the height of the hull deck, the gun mantlet hits the hatch covers when the gun is depressed. Therefore, the previous curved shape of the turret front must be maintained.

Item 8 - Elimination of the bulge under the commander's cupola is possible only if the turret side is sloped at an angle of 69 degrees instead of 60 degrees and also the commander's cupola is moved 50 mm farther over toward the turret centerline. An increase in weight of 400 kilograms will result if the turret side remains at 80 mm thick. However, if the plate thickness is increased from 80 to 90 mm because the angle is changed to 69 degrees, a further weight increase of 500 kilograms will result. The decision if and in which shape the bulge should remain is therefore urgent.

The need for design decisions in order for turret production to continue uninterrupted was discussed in another meeting between Oberstlt. Mueller of **Wa Pruef 6** and Krupp representatives on 15 January 1943:

About 20 armor bodies from the previous Tiger P2 turrets are already being machined; a further number are reputedly in the armor shop, so that about 40 to 50 turret armor bodies can be reckoned on shortly. The individual plates are already available for another 50 turrets. Therefore, a decision is urgently needed whether about 50 turrets of the previous model with 100 mm frontal armor should be produced to prevent a later gap in production or if production should be converted immediately to a turret model with increased frontal armor, with or without a side bulge.

On 29 January 1943, **WuG 2/IIId** informed Krupp that the **8.8 cm Kw.K.42** (drawing number 5-0808) now carries the designation **8.8 cm Kw.K.43 (L/71)** and drawing number 5-0808. This tank gun was designed by Krupp under contract SS 004-8104/41 awarded by **Wa Pruef 4/VIII**. A **geteilt** (sectional) gun tube was already produced for testing as **RV3**, the third **8.8 cm Kw.K.43**. The other six **8.8 cm Kw.K.43** (**RV1**, **RV2**, and **RV4** to **RV7**) produced under this contract all had **ungeteilt** (monobloc) gun tubes.

On 17 February 1943, Oberstlt. Crohn of **Wa Pruef 6** informed Krupp that OKH had approved the use of the 50 turrets that had already been forged. The reinforced Tiger H3 turret with sloped turret front would have to be used starting with the 51st Tiger H3.

Progress on the new turret design without a side bulge for the commander's cupola is shown in Krupp drawing 2 AKF 31702, which was incorporated into a Henschel drawing HSK Nr.3470 dated 3 June 1943. The flat turret front plate has been sloped at an angle of 10 degrees, the turret sides and rear at an angle of 20 degrees. Extending down to within 10 mm of the hull roof, the turret front plate and side plates provided protection for the turret ring at the front and sides. However, at the rear, the bottom of the turret overhang was elevated to 115 mm above the hull deck in order to clear the armor ventilator caps. Notches are shown cut out of the lower corners of the turret front plate to clear the armor ventilator cap and opened crew hatches on the hull roof. Lower edges of the turret sides had been beveled to eliminate overhang along the sides when the turret was positioned at 12 o'clock. The centerline for the **8.8 cm Kw.K.43** is shown to be 30 mm to the right of the vehicle and turret centerline. At 905 mm in diameter, the base of the commander's cupola extended 12.5 mm past the turret centerline. Details of the loader's hatch, rear hatch, ventilation fan, and gun mantlet had yet to be finalized.

The turret design for the **VK 45.03** that was finally approved by **Wa Pruef 6** for mass production by Krupp was assigned drawing number sequence 021 St 50600 through 50650. As revealed by the detailed drawing and parts list, practically every component was newly designed or modified for this turret. The cast armor commander's cupola with seven periscopes, originally designed for the Tiger P2 and then adopted for the Tiger I as 021 St 2762, was assigned to this Tiger II **"Serien-Turm."** As shown in Appendix C, many of the parts designed for the **"Serien-Turm"** were backfitted into the 50 turrets originally ordered for the Tiger P2 when these were modified for mounting on the first 50 **VK 45.03** chassis.

For further detailed descriptions, refer to Chapter 6.1 for the **"Turm Nr.1-50"** and Chapter 6.3 for the **"Serien-Turm."** Additional data on specifications, dimensions, and capabilities are provided as Tables in Appendix B.

6

Panzerkampfwagen Tiger Ausf.B

6.1 DESCRIPTION

The following detailed description and illustrations of the chassis designed by Henschel were extracted from **Firmen-Sammelmappe fuer Panzerkampfwagen Tiger Ausf.B** and **D656/43**, **Panzerkampfwagen Tiger Ausfuehrung B, Handbuch fuer den Panzerfahrer** dated 1 September 1944.

6.1.1 FAHRGESTELL (CHASSIS)

The main components of the chassis were the:
Panzerwanne (Armor Hull)
M.G. Kugelblende (Machinegun Ball Mount)
Winkelspiegel (Periscope)
Maybach HL 230 P30 Motor (Engine and Accessories)
Turmantrieb (Turret Drive)
Maybach OG 40 12 16 B Schaltgetriebe (Transmission)
L 801 Zweiradienlenkgetriebe (Steering Gear)
Bremsen (Brakes)
Seitenvorgelege (Final Drives)
Gestaffeltes Laufwerk (Overlapping Suspension)
Schwingungsdaempfer (Shock Absorbers)
Gleisketten (Tracks)
Leitrad (Idler Wheel)
Elektrische Anlage (Electrical System)
Selbsttaetige Feuerloescheinrichtung (Automatic Fire Extinguisher System)
Durchdrehvorrichtung (Hand-Cranked Starter)
Kurbelwellen-Benzinanlasser (Gasoline Engine Starter)
Kraftstoffbehaelter (Fuel Tanks)
Munitionslagerung fuer Kw.K. (Ammunition Stowage)
Halterungen mit Zubehoer aussen am Pz.Kpfw. (External Stowage)

6.1.1.1 Panzerwanne (Armor Hull)

The new hull design consisted of sloping plates for increased protection. The front glacis plate was 150 mm at 50 degrees, front nose plate 100 mm at 50 degrees, superstructure side plates 80 mm at 25 degrees, hull side plates 80 mm at 0 degrees vertical, tail plate 80 mm at 30 degrees, deck plates 40 mm at 90 degrees horizontal, and front belly plate 40 mm horizontal and rear belly plate 25 mm horizontal. A large rectangular cover plate, flush with the roof in front of the turret, could be lifted to remove the transmission and steering units for maintenance without having to remove the turret. Hatches for the driver and radio operator were cut into this cover plate. There was a large, hinged rectangular hatch over the motor, and the entire rear deck could be removed for maintenance on the motor, cooling system, and fuel system.

6.1.1.2 M.G. Kugelblende (Machinegun Ball Mount)

A cast armor guard for protecting the machinegun ball mount was welded into a circular opening in the glacis plate. The aperture in front of the ball mount was stepped to avoid bullets being deflected into the ball joint. Internally the ball joint was protected against bullet splash. The ball-mounted **M.G.34** was manned by the radio operator, who aimed it with the **Kugelzielfernrohr 2** (sighting telescope).

6.1.1.3 Winkelspiegel (Periscope)

The periscope for the driver was mounted in a revolving socket which permitted vision in any direction. The periscope for the radio operator, however, was stationary at a position of 16.5 degrees to the right front. The part of the periscope which protruded through the hull and revolving plate was protected against shattering.

The upper prism glass was mounted in a flanged socket; the lower one could be moved sideways. Both periscopes were embedded in rubber seals and could be removed by pulling them sideward.

A slit with a carrier to hold the driver's periscope was fitted into the revolving socket in the hull roof. Circular guides with hand grips allowed the entire periscope to swing round its vertical axis

Panzerkampfwagen Tiger Ausf.B - Fgst.Nr.V1 completed in November 1943 – hinged flat track guards on the front – no tool stowage or **Zimmerit** (as in most **Versuchs-Serie**) – 25 mm thick plates fore and aft on the turret roof (as produced originally for the **VK 45.02 (P)**.

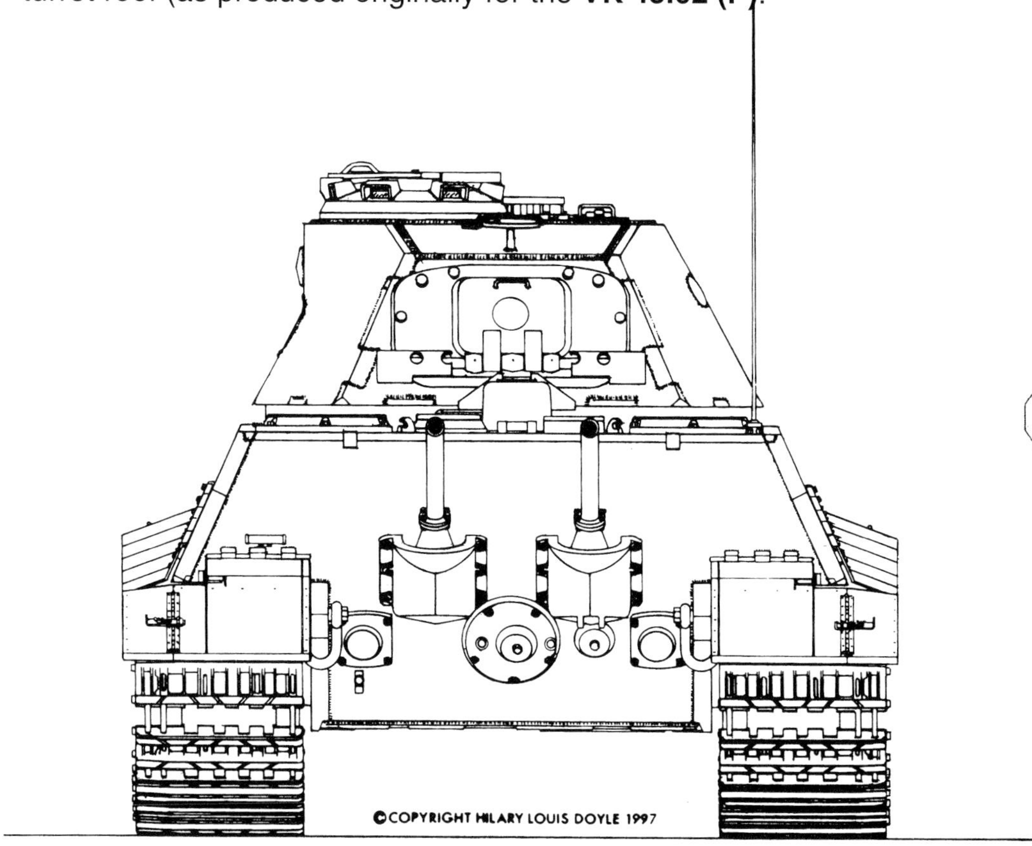

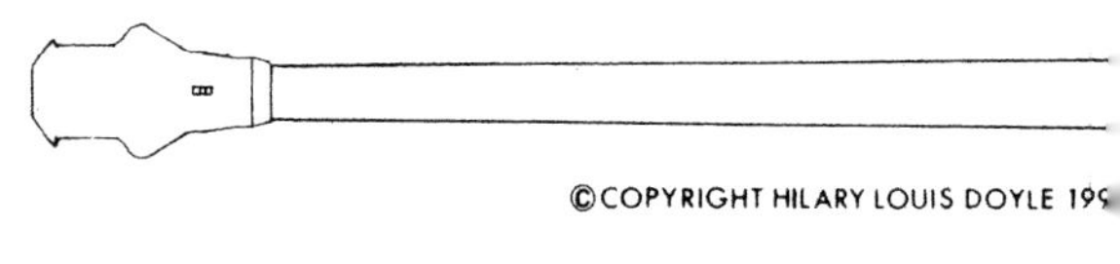

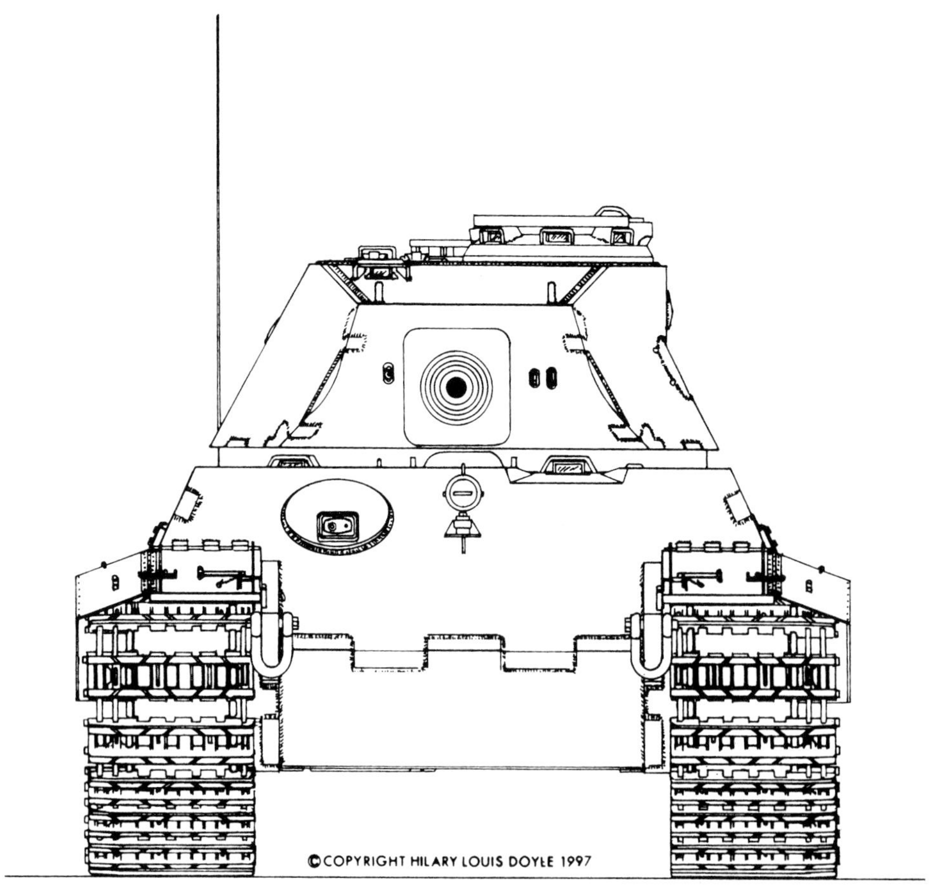

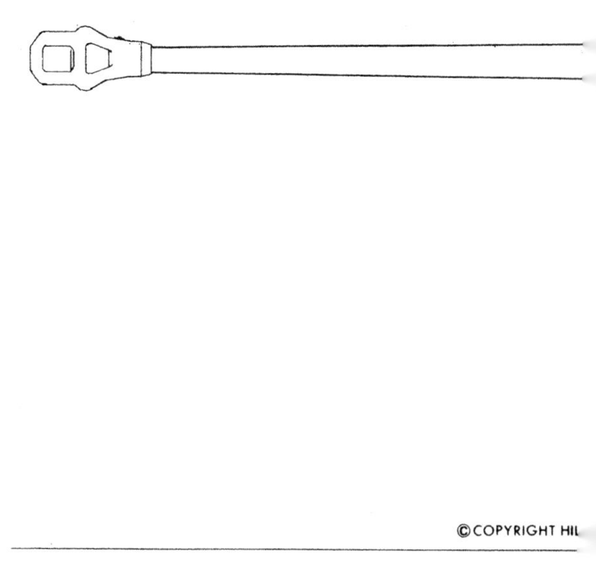

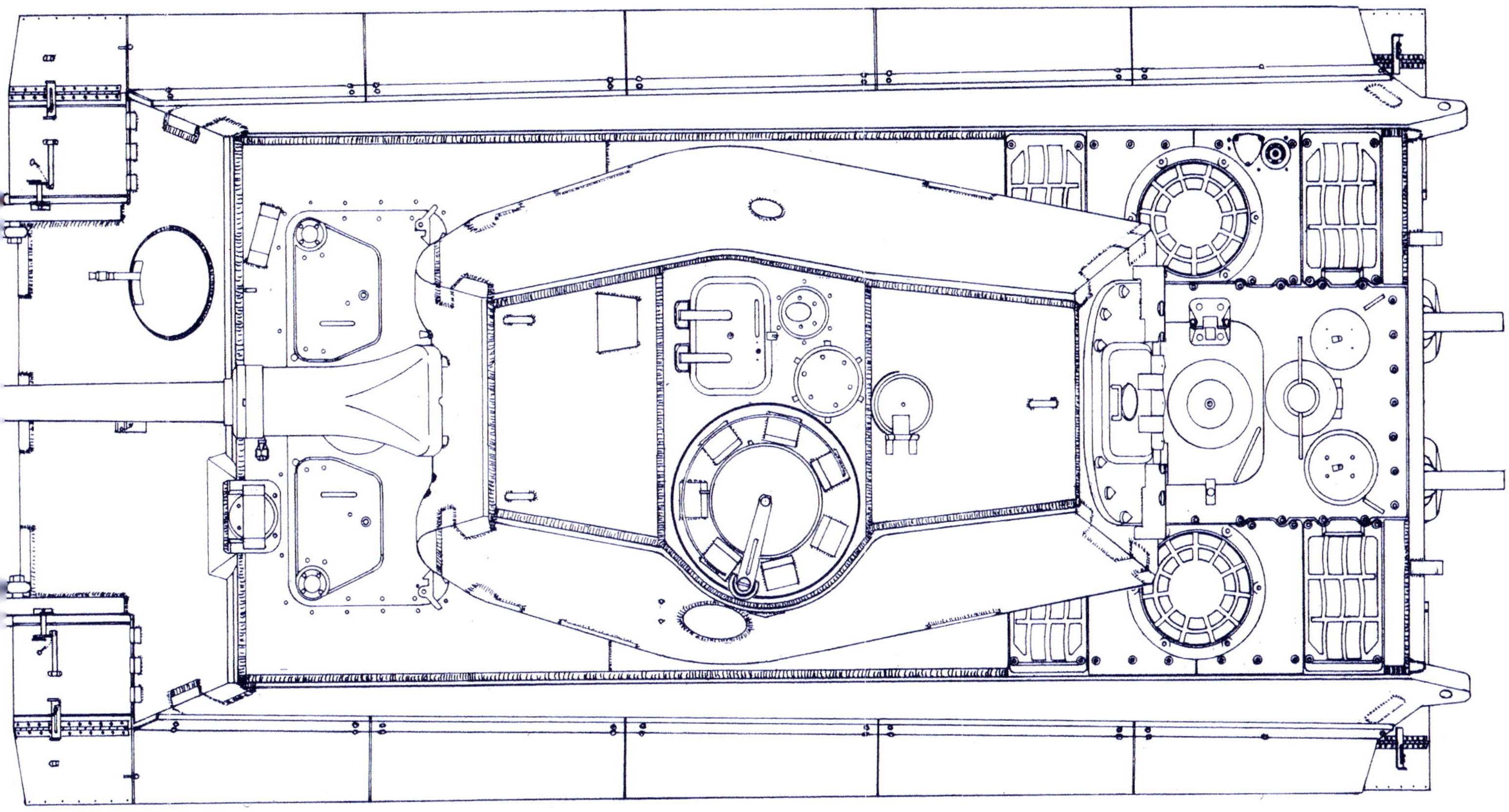

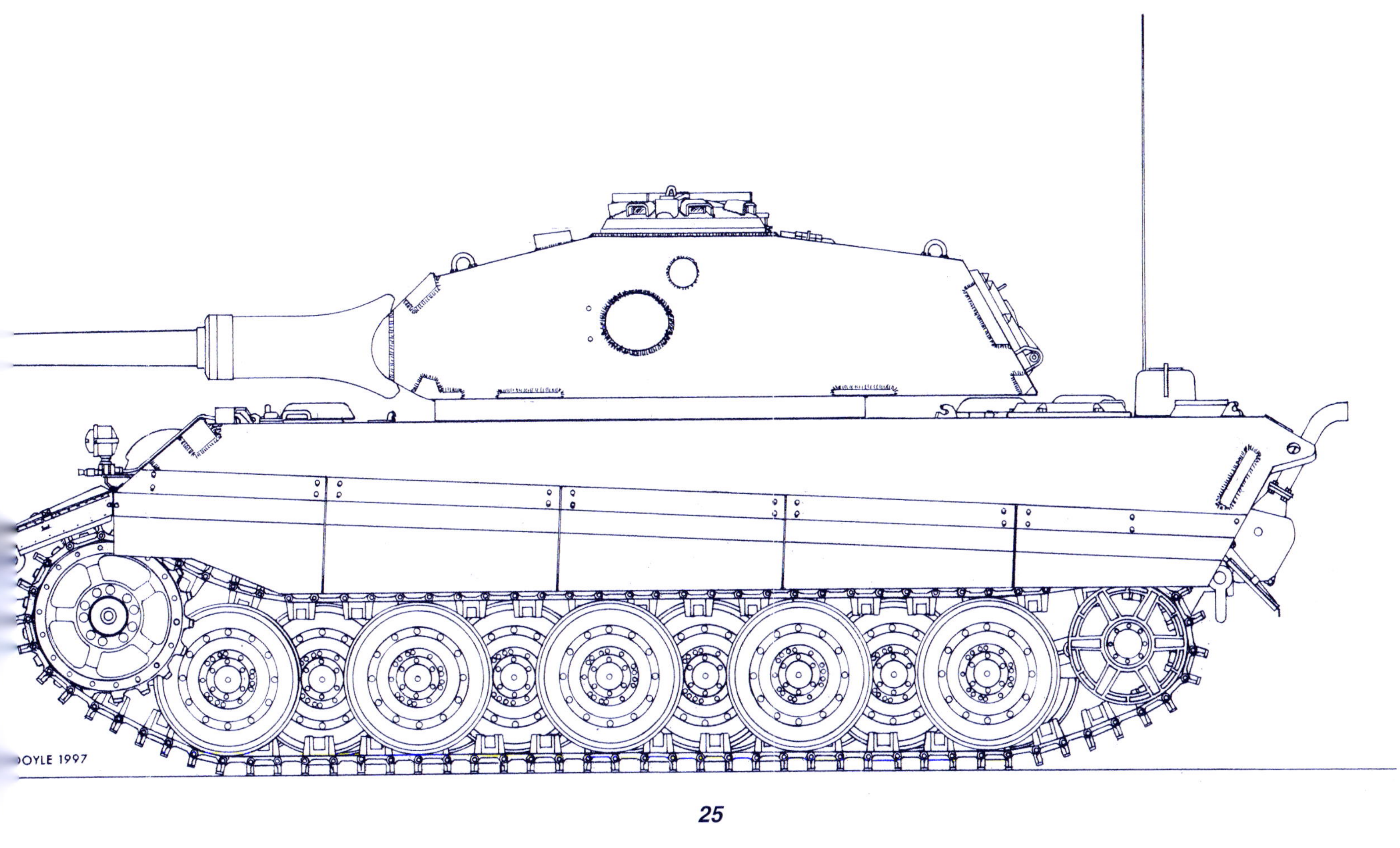
DOYLE 1997

and, at the same time, it could be tilted horizontally. The revolving plate was protected against water leaking in by two rubber seals and a grease-filled groove.

The driver's seat, steering controls and accelerator pedal were adjustable to allow freedom and ease of driving with the driver's head sticking out of the open hatch.

6.1.1.4 Maybach HL 230 P30 (Engine and Accessories)

Power for the Tiger II was provided by a **Maybach HL 230 P30**, 12 cylinder, 23 liter (145 mm stroke, 130 mm bore) engine with a compression ratio of 6.8 to 1, rated at 600 horsepower at 2600 rpm.

The **HL 230 P30** engine was equipped with a governor which performed the following functions: 1) activating the second stage of the carburetor and 2) regulating the engine speed.

When the engine exceeded 2000 rpm, the outer spring in the centrifugal governor released a slide in the oil feed. This brought a piston into operation. The piston was linked to a second slide which could be controlled by the gas pedal. The second stage of the carburetor was only engaged if: a) the revolutions exceeded 2000 rpm, or b) the gas pedal was pushed all the way down.

As soon as maximum revolutions were reached, the inner spring of the centrifugal governor closed the intermediate chokes in the suction pipe. Engine speed was limited quite independently of the position of the carburetor choke (butterfly valve).

[The driver's handbook, dated 1 September 1944, stated that the motor was governed at a maximum speed of 2500 rpm and that the second stage of the carburetors could be activated at an engine speed exceeding 1600 rpm.]

Air was drawn though the cyclone air filters mounted on top of the carburetors. Dust particles, removed by centrifugal action, dropped into two dust trap compartments on both sides of the engine. Then the cleaned air passed through the oil bath, which removed any remaining dirt. The filters were to be cleaned whenever refueling occurred but did not have to be cleaned more frequently even in extremely sandy regions.

Two mechanically driven fuel pumps which were located on the left side of the motor housing pumped fuel from the main fuel line into a pipe which led to the carburetor and filled the float chamber. The fuel line was routed in such a way that its highest point was above the level of the fuel in the fuel tank. An air vent system for the fuel tank prevented a siphon effect in the suction pipe.

The fuel pumps were of the membrane type, driven by a push rod. In case the float chamber was full, the fuel pump idled. The HL 230 P30 engine was not equipped with an electrical fuel pump but had a small hand pump with which it was possible to pump fuel into the carburetor avoiding the mechanical pumps. This could be done after opening a flap in the firewall.

The engine compartment was ventilated by radiator fans drawing air out through a duct in the left bulkhead.

The engine exhaust manifolds were cooled by a fan located behind the engine which drew air in through a duct connected to a vent on the deck of the tank. The air was pushed forward through ductwork to the firewall and from there into the two separate ducts, one for each side, connected to the mantles surrounding the exhaust headers. After cooling the exhaust, the air was drawn out by the radiator fans through ducts penetrating the engine compartment side bulkhead.

The exhaust lines from the engine were connected to the tail pipes on the hull rear through a double ball and socket joint so as to not interfere with engine movement. The penetrations for the two exhaust pipes in the rear were protected by heavy cast armor shields, and the tail pipes were surrounded by sheet metal shields.

6.1.1.5 Turmantrieb (Turret Drive)

The turret drive housing incorporated intermediate bearings for the split drive shaft between the engine and the transmission. The turret traverse was powered by the main driveshaft being connected through a reduction gear to the hydraulic drive mounted on the turret platform. The turret drive idled until engaged through the hydraulic drive. The turret could be traversed at two different speeds selected by using a hand lever connected to the hydraulic drive. This enabled coarse laying of the gun; fine laying was done by the gunner using a handwheel.

6.1.1.6 Maybach OG 40 12 16 B Schaltgetriebe (Transmission)

The pre-selective shifted transmission had eight forward and four reverse speeds. Maximum speed for each gear was limited by the maximum engine speed.

Forward	km/hr at 3000 rpm	km/hr at 2500 rpm
1st Gear	2.54	2.12
2nd Gear	3.83	3.19
3rd Gear	5.62	4.68
4th Gear	8.33	6.94
5th Gear	12.75	10.62
6th Gear	18.95	15.79
7th Gear	27.32	22.77
8th Gear	41.5	34.6
Reverse		
1st Gear	3.39	2.82
2nd Gear	5.11	4.26
3rd Gear	7.50	6.25
4th Gear	11.12	9.27

The transmission oil was cooled in a flexible pipe which led through a water-filled cooler installed behind the transmission in the fighting compartment.

6.1.1.7 L 801 Zweiradienlenkgetriebe (Steering Gear)

The L 801 double-radius steering gear was designed to allow the shortest turning radius of 2.08 meters in first gear and the largest turning radius of 114 meters in eighth gear.

6.1.1.8 Bremsen (Brakes)

The disc brakes were engaged either by brake pedal or brake lever. The latter could be locked (emergency brake). The brakes were only to be used for steering purposes in an emergency.

The brake housing rotated; the brake lining was held only

loosely in contact with the brake discs. The process of braking was mechanical. The actual braking effect was exerted by 29 ball bearings which were located between the brake discs. Tightening of the discs was effected with the help of six springs. After the brake lining had been worn, the linkage of the brake lever had to be shortened. The width of the brakes (disc and lining) was 66 mm. Brake clearance could be measured through small holes in the brake housing on each side. If the width had been worn to less than 58 mm, it had to be replaced.

6.1.1.9 Seitenvorgelege (Final Drives)

The final drive housing, with a heavy central flange supporting the brake carrier, was mounted in a hole cut into the hull side. Power was transmitted from the steering gear through two drive shafts, one to each final drive on the right and left side. Each drive shaft had two double herring bone reduction gears which fitted into two similar gears, insensitive to small deviations. Both final drive housings were symmetrical and interchangeable.

The side shafts of the steering gear were connected by flanges to the pinion on the final drive shaft. Reduction of revolutions was effected first by a large spur gear and then by a planetary gear system. The planetary gear formed the reduction shaft to the drive sprocket wheel. The bell-shaped drive sprocket wheel was a heavy armor casting, secured with ten screws and eight bolts.

6.1.1.10 Gestaffeltes Laufwerk (Overlapping Suspension)

The overlapping suspension consisted of eight inner and ten outer pairs of roadwheels. They were arranged as five outer and four inner pairs of roadwheels on each side. Each roadwheel suspension arm was connected to a torsion bar inside the hull through a lubricated plastic bushing. Torsion bars on the first and last wheel arm were stronger, but all had the same type of splined ends.

Roadwheels were made of strong sheet steel discs sandwiching two rubber rings which in turn held the outer steel rim. The roadwheels were fastened to the hub with help of two conical rings which were divided and held together by a thumb screw.

Hubs for inner and outer roadwheels were different, but both had tapered roller bearings. Attention had to be paid to exact adjustment to avoid over-tightening the bearings. If roadwheels were removed, the hub was to remain on the suspension arms.

6.1.1.11 Schwingungsdaempfer (Shock Absorbers)

The forward shock absorbers were connected to the two forward roadwheel suspension arms on each side, beside the driver and radio operator. The two rear shock absorbers were connected to the two rear roadwheel suspension arms and were located inside by the lower fuel tanks. They absorbed shock only on the downward stroke which was limited by a rubber bump stop struck by the suspension arm.

6.1.1.12 Gleisketten (Tracks)

Without sinking-in, ground pressure was 1.02 kg/cm2 with the 800 mm wide **Gelaendekette** (cross-country track) compared to 1.23 kg/cm^2 with the 660 mm wide **Verladekette** (transport track). The cross-country track was too wide for the railroad profile and therefore had to be replaced by the transport track when the Tiger II was shipped by rail. Weight of a double link cross-country track was 62.7 kg, transport track 42.9 kg.

The cross-country track had 46 double links which were connected by track pins. These track pins had a head on the inside and a locking ring on the outside. The locking ring was secured by a retaining pin fitted through a hole in the track pin. Each double track link consisted of one bridge link, one connecting link, three side links (as auxiliary links in the form of connecting bars), and two track pins.

The transport track was also used as the cross-country track for the Panther II. Instead of locking rings, the transport track had split rings to secure the track pins.

Grousers could be added to improve the grip of the track.

6.1.1.13 Leitrad (Idler Wheel)

The rear idler was mounted on two sets of ball bearings on the idler crank. The idler crank was mounted on two sets of lubricated plastic bearings, one in the hull side and one in the lower hull partition. The track tensioner was adjusted by using a double box wrench.

6.1.1.14 Elektrische Anlage (Electrical System)

The entire electrical system consisted of ignition, generator, starter, batteries, and battery heater.

Generator - With the engine running, the generator supplied all power take-off points and also charged the batteries. The voltage output was maintained independent of engine speed by a regulator. The regulator also governed the generator output in relation to the condition of the batteries to avoid overcharging. The commutator was equipped with a fan to prevent overheating. The regulator was mounted separately from the generator unit. It also had an automatic switch which made a parallel connection between the generator and the batteries as soon as the revolutions were high enough to produce the required current to charge the batteries.

Starter Mechanism - The Bosch starter had a primary electric motor with adjustable armature. A switch system which worked in two voltage steps engaged the starter when the starter button was pressed. Armature shaft and pinion were connected through a friction clutch, which permitted the shaft to idle as soon as the engine started. The starter required 24 volts.

Switch System - Two batteries of 12 volts each were connected in parallel. If the starter button was pushed, a battery switch connected both batteries in series and increased the voltage to 24 volts.

Battery - The master switch which disconnects the entire electrical system was located on the firewall and could be reached from the fighting compartment. All lines were fuzed. On the panel were two fuze boxes with eight 15 amp fuzes and two 40 amp fuzes. The same panel was equipped with a junction box for plugging in an extension hand-held light. All lights were 12 volts.

Battery Heaters - At very low outside temperatures, the batteries could be heated by either a 100 watt or a 300 watt heater. The 100 watt plate served as a heating unit when on the move.

The 300 watt heater could be connected to an external power source at a plate on the firewall. The batteries could also be recharged by an external power source connected through this plate.

6.1.1.15 Selbsttaetige Feuerloescheinrichtung (Automatic Fire Extinguisher System)

The fire extinguisher system was installed with its heads aimed to put out fires at the carburetors and fuel pumps. It used "CB" as the extinguishing agent and sprayed through four nozzles. The three liters of fluid were sufficient for five discharges. Heat sensing heads were connected with the nozzles. When temperatures reached 120oC, the valves opened and discharged the fluid for a period of 7 seconds. If the fire was not extinguished by then, the timer reopened the valves for another 7 second period and so on, rewinding itself during each extinguishing period.

6.1.1.16 Durchdrehvorrichtung (Hand-Cranked Starter)

A reduction gear starter was installed just above the electrical starter. When the engine was not running, the starter linkage was used to push a pinion through until it fit into the flywheelgear. In this way the flywheel could be turned by the crew using a handcrank. As soon as the engine started, the pinion disengaged itself.

The handcrank fitted into a small opening on the hull rear protected by an armor cover just below the right-hand exhaust pipe. The handcrank shaft was supported by a ring pivoting in the two eyelets fitted to the underside of the armor guard for the right-hand exhaust pipe.

6.1.1.17 Kurbelwellen-Benzinanlasser (Gasoline Engine Starter)

Directly behind the middle of the engine crankshaft was a large opening in the hull rear. This was protected by an armor cover upon which were welded two pegs for mounting the gasoline engine starter. The large armor cover remained in place at all times. The connecting shaft for the starter was inserted through a smaller hole centered in the large cover. The starter was secured to and rested on both palls.

6.1.1.18 Kraftstoffbehaelter (Fuel Tanks)

The 860 liters of fuel were stored in seven tanks as follows:

- One fuel tank with filling tube in upper rear side in motor compartment, capacity approximately 85 liters
- Two fuel tanks on upper left and right side of motor compartment, which were submerged when the Panzer crossed water, capacity approximately 290 liters
- Two fuel tanks at left and right of bottom in motor compartment, capacity right tank 65 liters, left tank 80 liters
- Two fuel tanks at left and right side in fighting compartment underneath ammunition boxes, capacity approximately 340 liters.

All fuel tanks were connected in such a manner that they could all be filled through the one opening located on the rear deck of the engine compartment. A battery of valves made it possible to empty all the tanks in three phases. All fuel tanks could be emptied through a drain plug in the hull floor. All tanks were also interconnected with an air vent system.

6.1.1.19 Munitionslagerung fuer Kw.K. (Ammunition Stowage)

The ammunition bins were fitted with metal sliding panels which were to be kept closed at all times. Forty-eight rounds were stowed in the fighting compartment. They could be either H.E. or A.P. Ammunition racks were fitted along the panniers in both hull sides. Six rounds were stored in the front racks, seven in the middle, and eleven at the rear.

6.1.1.20 Halterungen mit Zubehoer aussen am Pz.Kpfw. (External Stowage)

On the rear:
1 20-ton steel jack with lifting housing and folding crank
1 wooden block for the 20-ton jack
2 C-hooks
On the left side:
1 32 mm diameter steel cable, 8.2 meters long
1 wrecking bar, 1200 mm long
1 handcrank for the starter
1 spade
3 cleaning rods
On the right side:
1 32 mm diameter steel cable, 8.2 meters long
1 14 mm diameter steel cable, 15 meters long
3 cleaning rods
On the deck:
1 2-liter Tetra fire extinguisher
1 axe
1 sledge hammer
1 wire cutter

The first **Pz.Kpfw.Tiger Ausf.B** (**Fgst.Nr.V1**, completed by Henschel in November 1943) had several features unique to the **Versuchs-Serie** (trial series), including the lack of **Zimmerit** (anti-magnetic coating), lack of tool stowage, and flat front mudguards with hinged side extensions. Both the **Schauloch** (view port) and **MP-Stopfen** (pistol port) on the left side of the turret have been welded shut. (WJS)

Each double link of the 660 mm wide **Verladekette** (transport track) was made up of a bridge link with eight chevrons, a connecting link, one side link, and two track pins. (WJS)

The 800 mm wide **Gelaendekette** (cross-country track) was made up with 46 bridge links and 46 connecting links. Each double link consisted of a bridge link with eight chevrons, a connecting link, three side links, and two track pins. (BA)

Originally the drive sprocket ring had 18 teeth to engage both the bridge link and the connecting link of the double link track. The internal rubber rings cushioning the steel-tired roadwheels were torqued down by 12 bolts on the wheel discs. (WJS)

As shown in this photo of **Tiger Ausf.B** (**Fgst.Nr.V1**) taken 19 December 1943, there was very little clearance between the hub of the six-spoke idler wheel and the last roadwheel on the right side. (WJS)

BELOW: The drive sprocket wheel was used as a capstan to pull on the cable connected to the **Gelaendekette** (cross-country track), as shown in this photo of **Tiger Ausf.B** (**Fgst.Nr.V1**) changing tracks on 23 December 1943. (WJS)

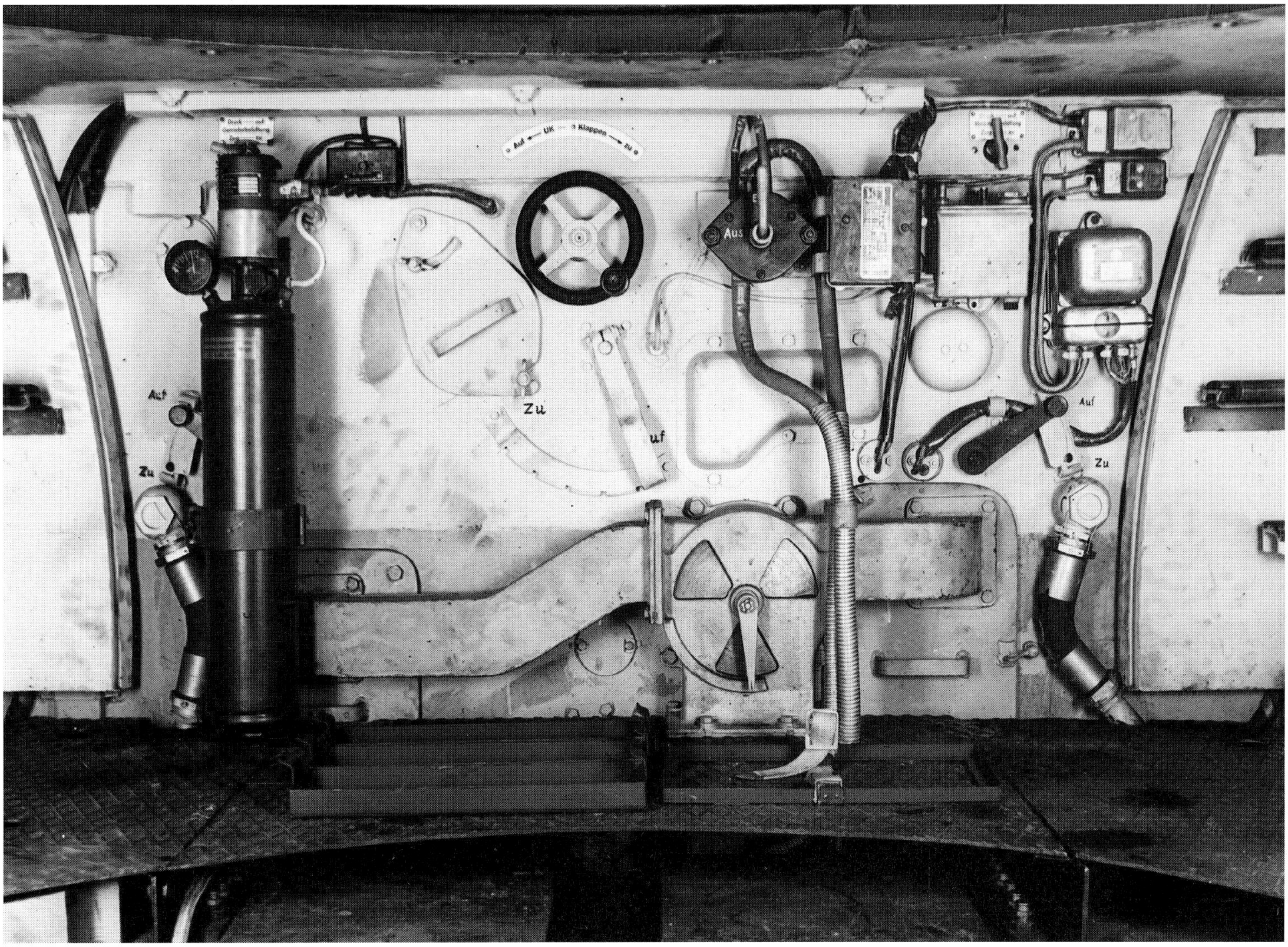

ABOVE: The unique feature on the fire wall of **Tiger Ausf.B** (**Fgst.Nr.V2**, completed by Henschel in January 1944) was the handwheel in the upper center. The handwheel was used for opening and closing vents to allow "**UK**" (underwater travel). Other equipment mounted on the firewall included the automatic fire extinguisher system, the master switch for the electrical system, other controls for operating dampers for the radiators, and the ductwork for cooling the engine exhaust manifold. (WJS)

LEFT: The rear of **Tiger Ausf.B** (**Fgst.Nr.V2**) showing that it still had the original **Gg 24/800/300** cross-country tracks fitted when originally picked up by the British in Haustenbeck, Germany. The hinged armor cap on the rear deck protected the telescoping air intake tube when it was retracted. The exhaust pipes were modified by Henschel for testing exhaust pressure. (NA)

Now on display at The Tank Museum in Bovington, England, **Pz.Kpfw.Tiger Ausf.B** (**Fgst.Nr.V2**) was acquired by the British at the Henschel testing area in Haustenbeck, Germany at the end of the war. It had been outfitted with standard issue tools and equipment fastened to the hull but wasn't covered with **Zimmerit** (anti-magnetic coating). (TTM)

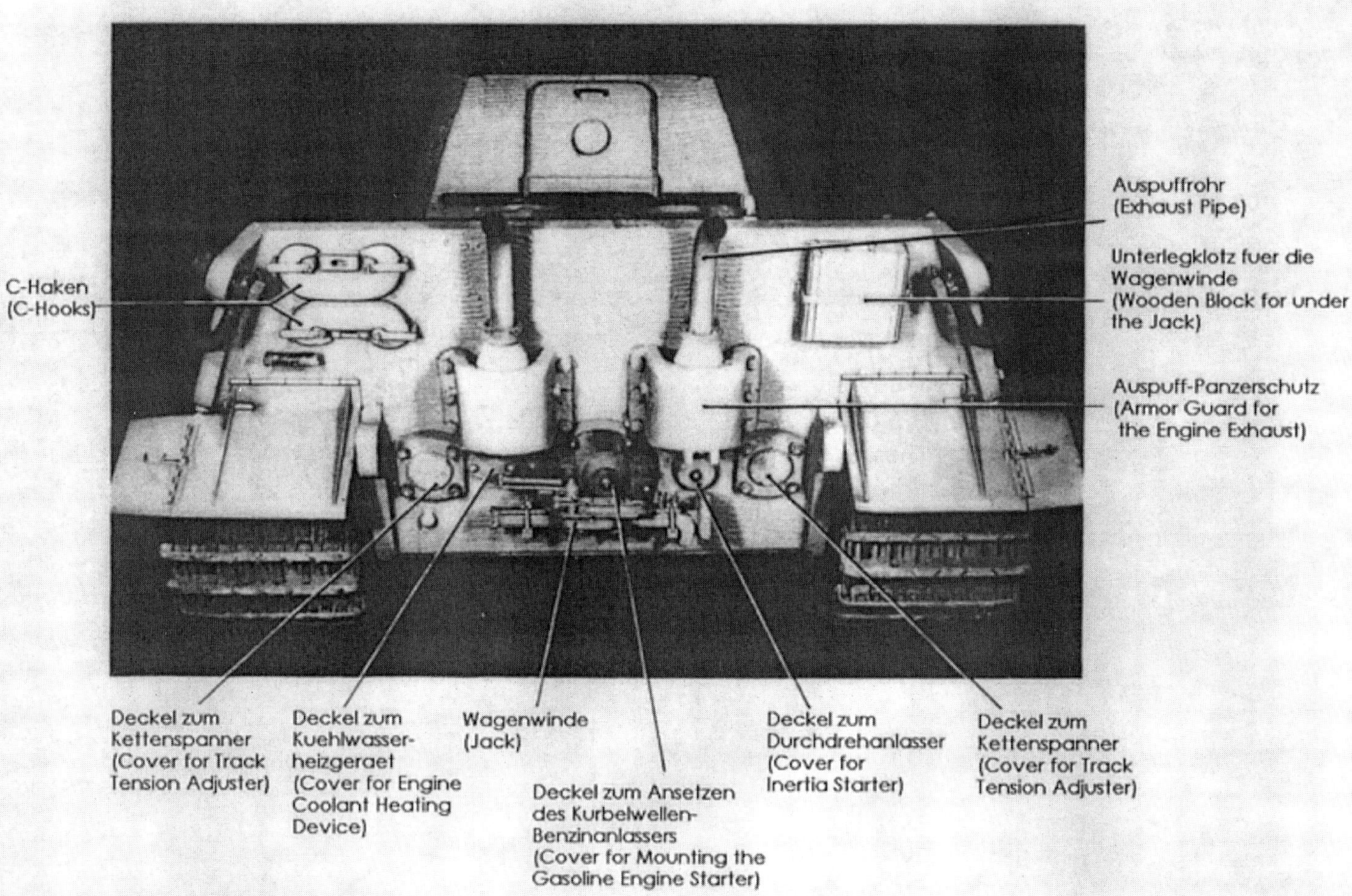

The rear of a Tiger II with a **Serienturm** showing the tool stowage arrangement and location of openings. The **Kuehlwasserheizgeraet** (engine coolant heater) was first installed in February 1944 and the **Serienturm** (series turret) was first installed in June 1944.

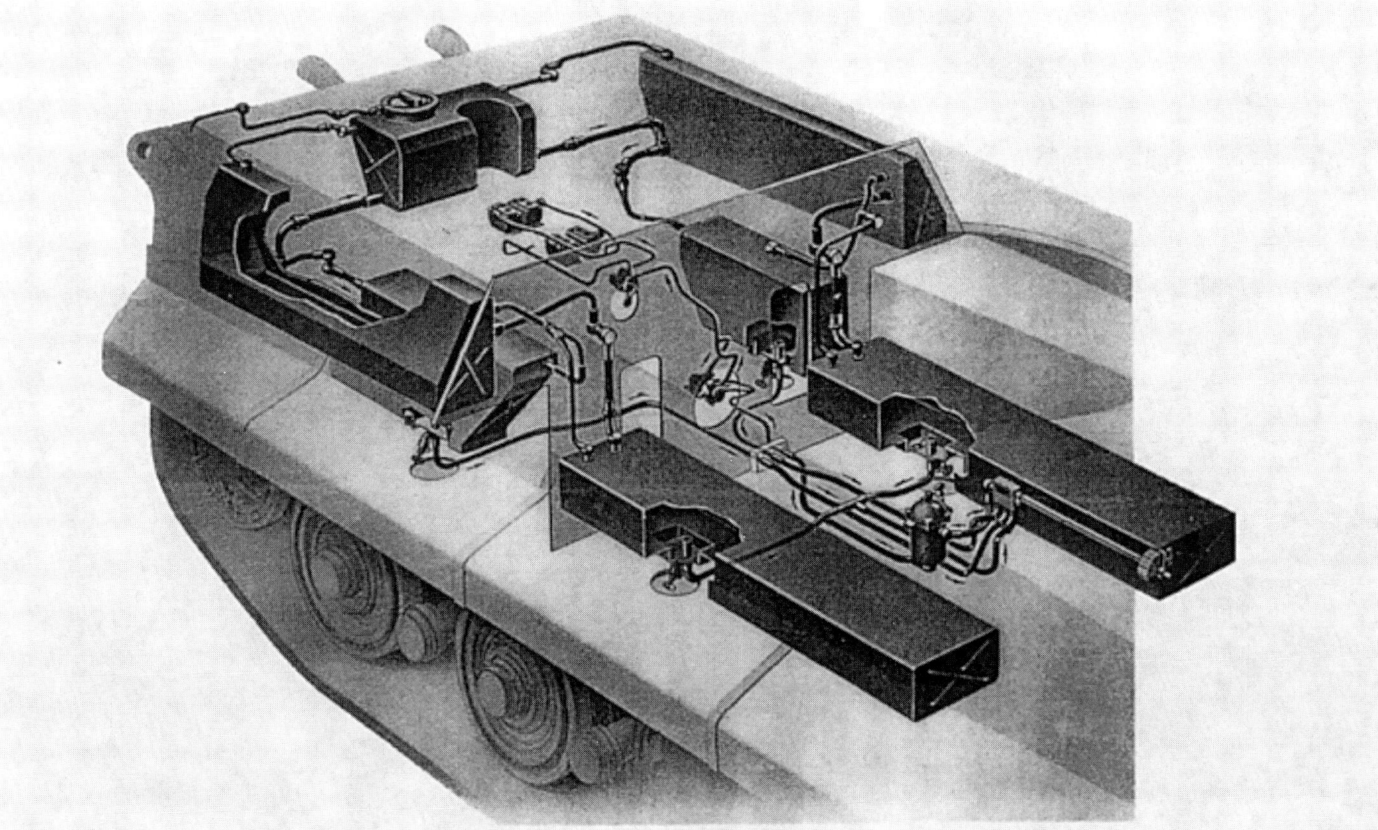

Kraftstoffbehaelter (fuel tanks) with connecting lines, pumps, and valves. Leaks from the numerous connections (about 180) caused fires in the engine compartment.

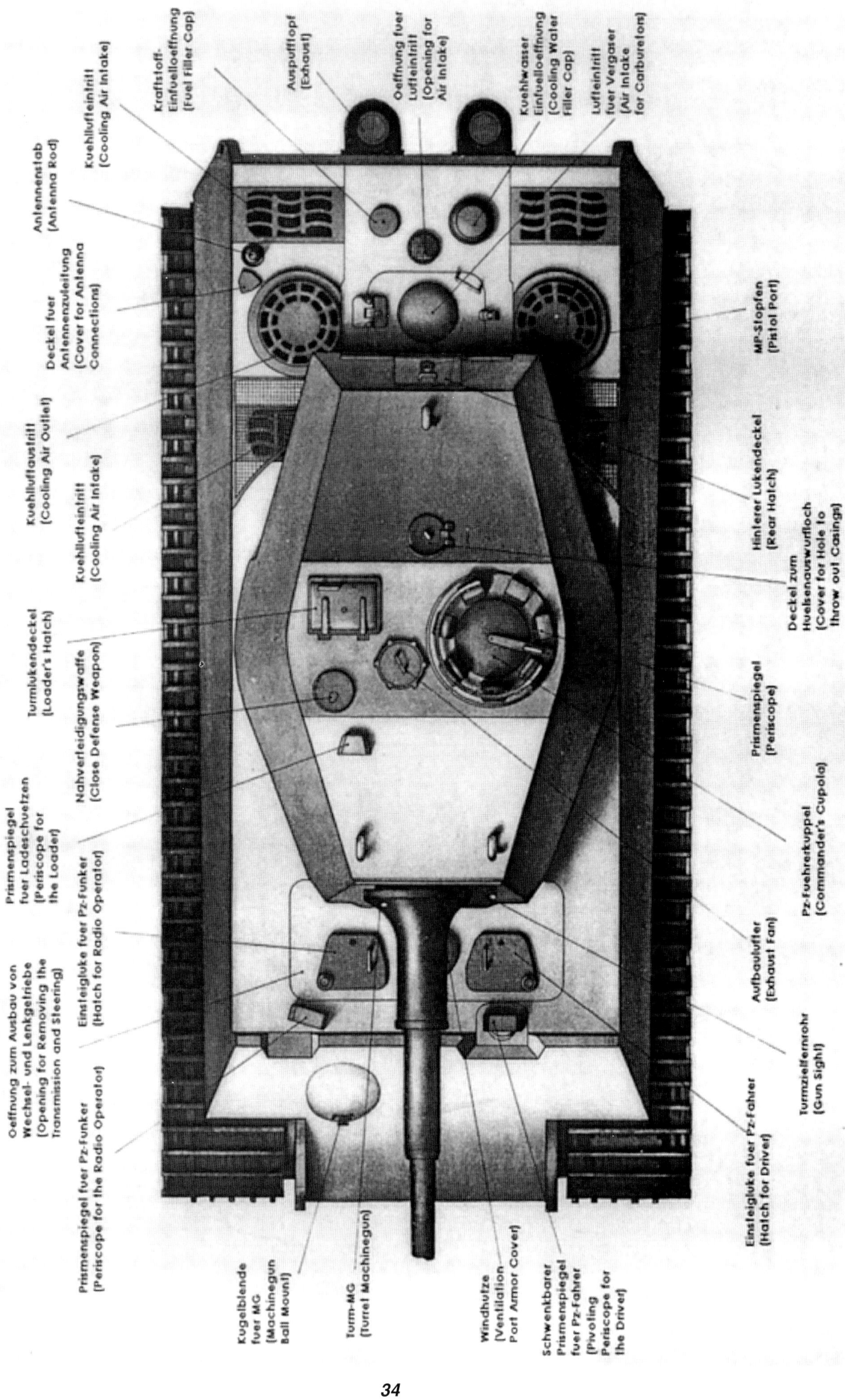

Location of components on the Tiger II (Not to Scale)

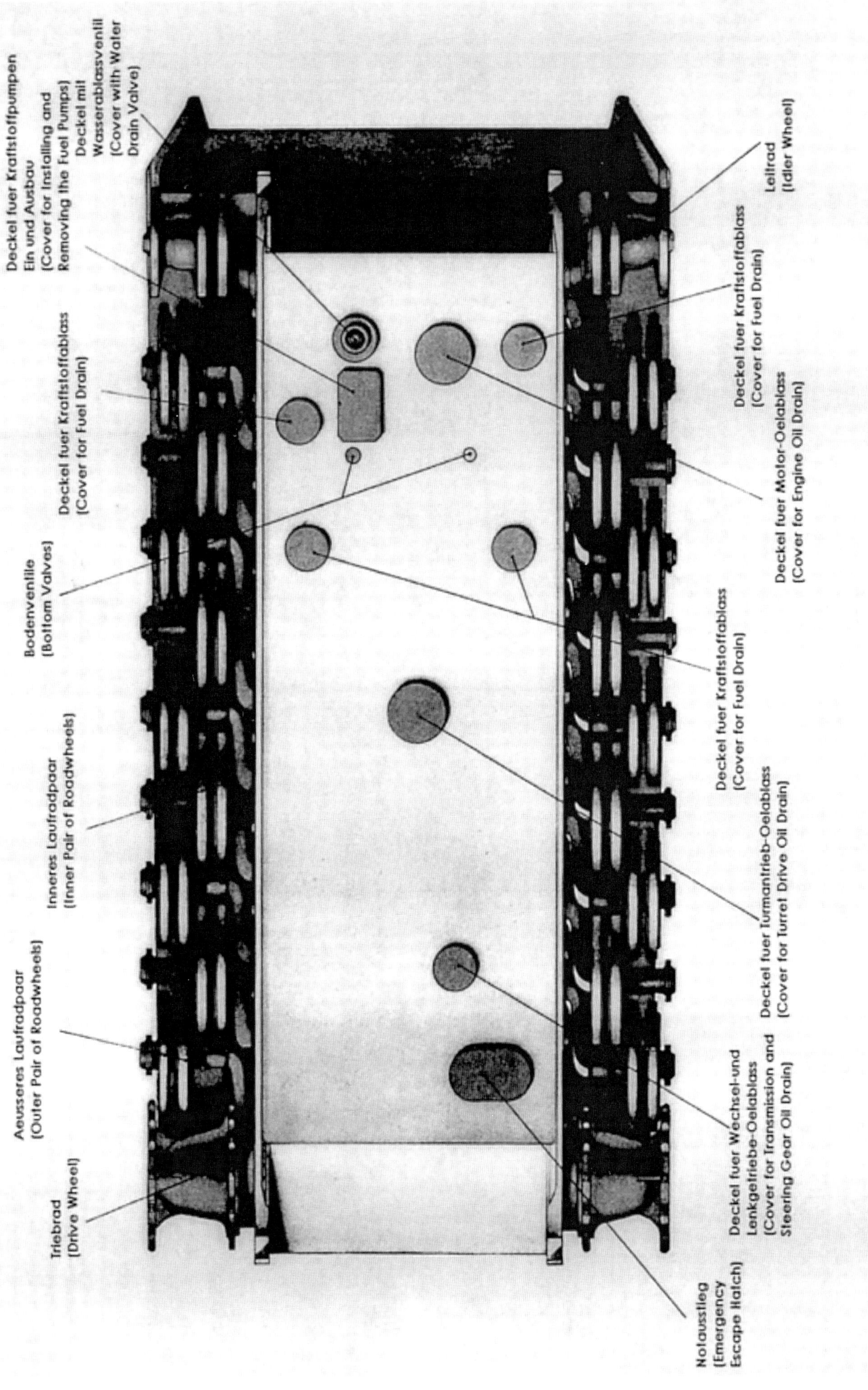

Location of openings in the belly of the Tiger II (Not to Scale)

Arrangement of drive train components in the Tiger II (Not to Scale)

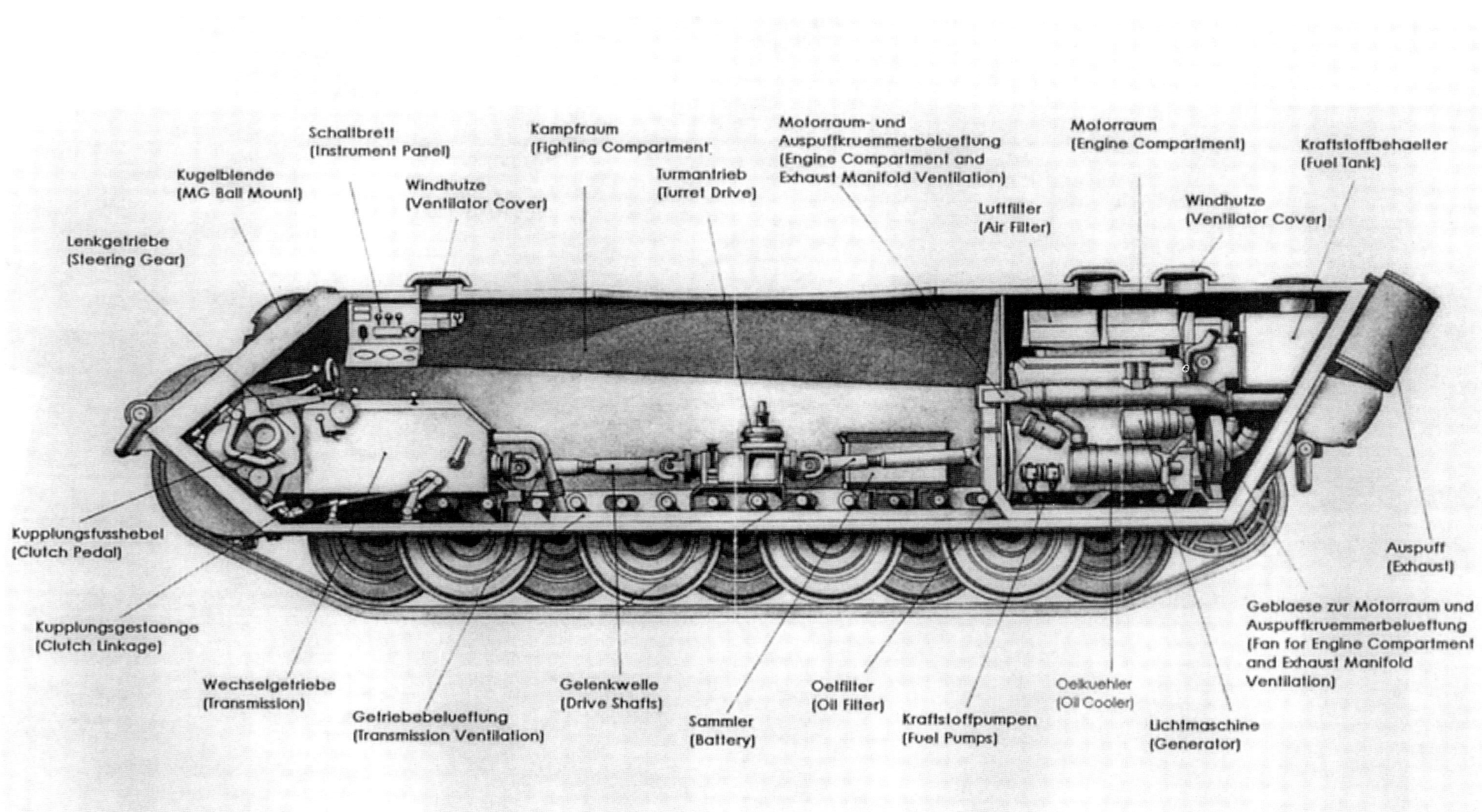

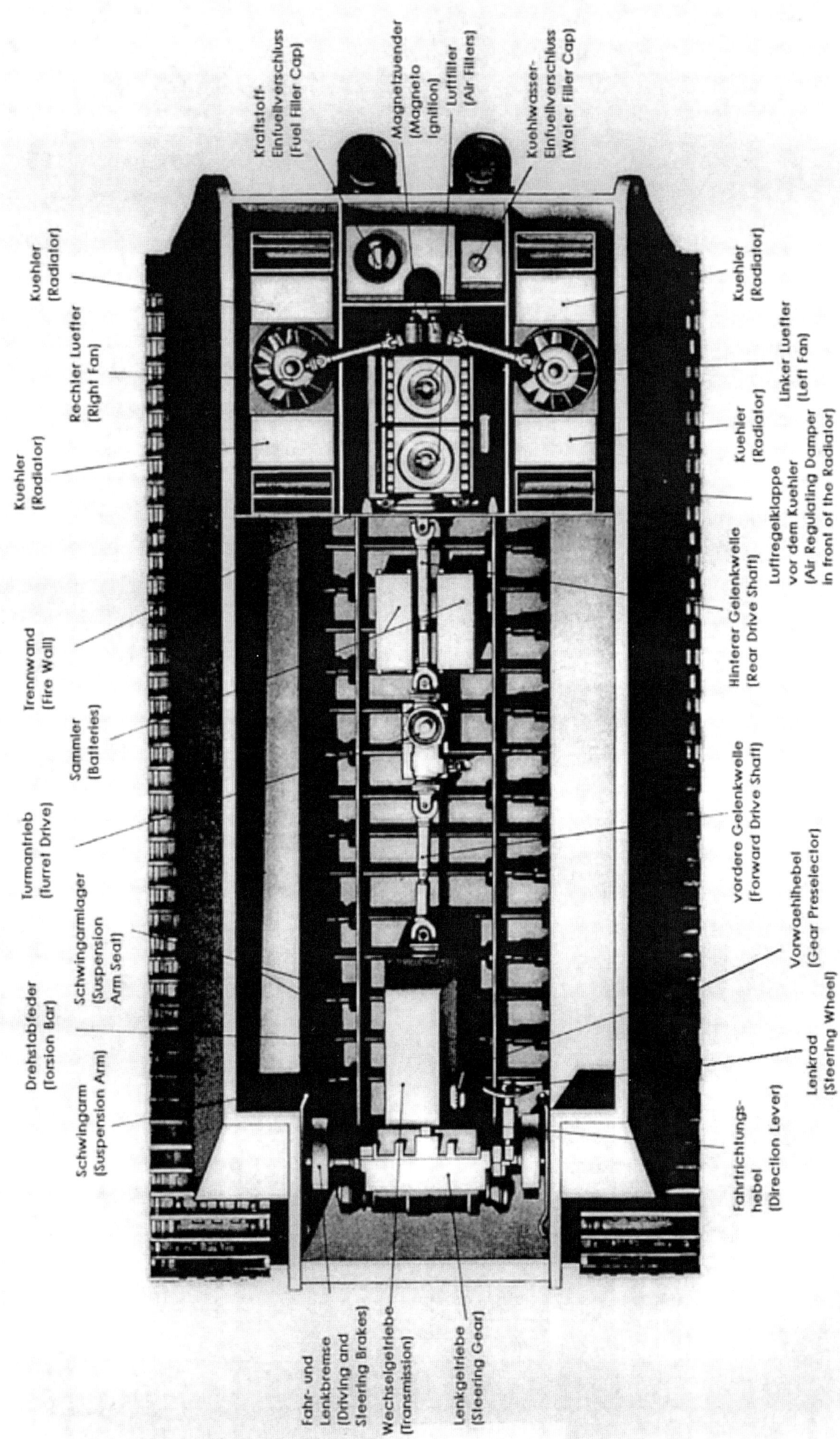

Arrangement of drive train components in the Tiger II (Not to Scale)

Arrangement of components inside the front of the Tiger II chassis (Not to Scale)

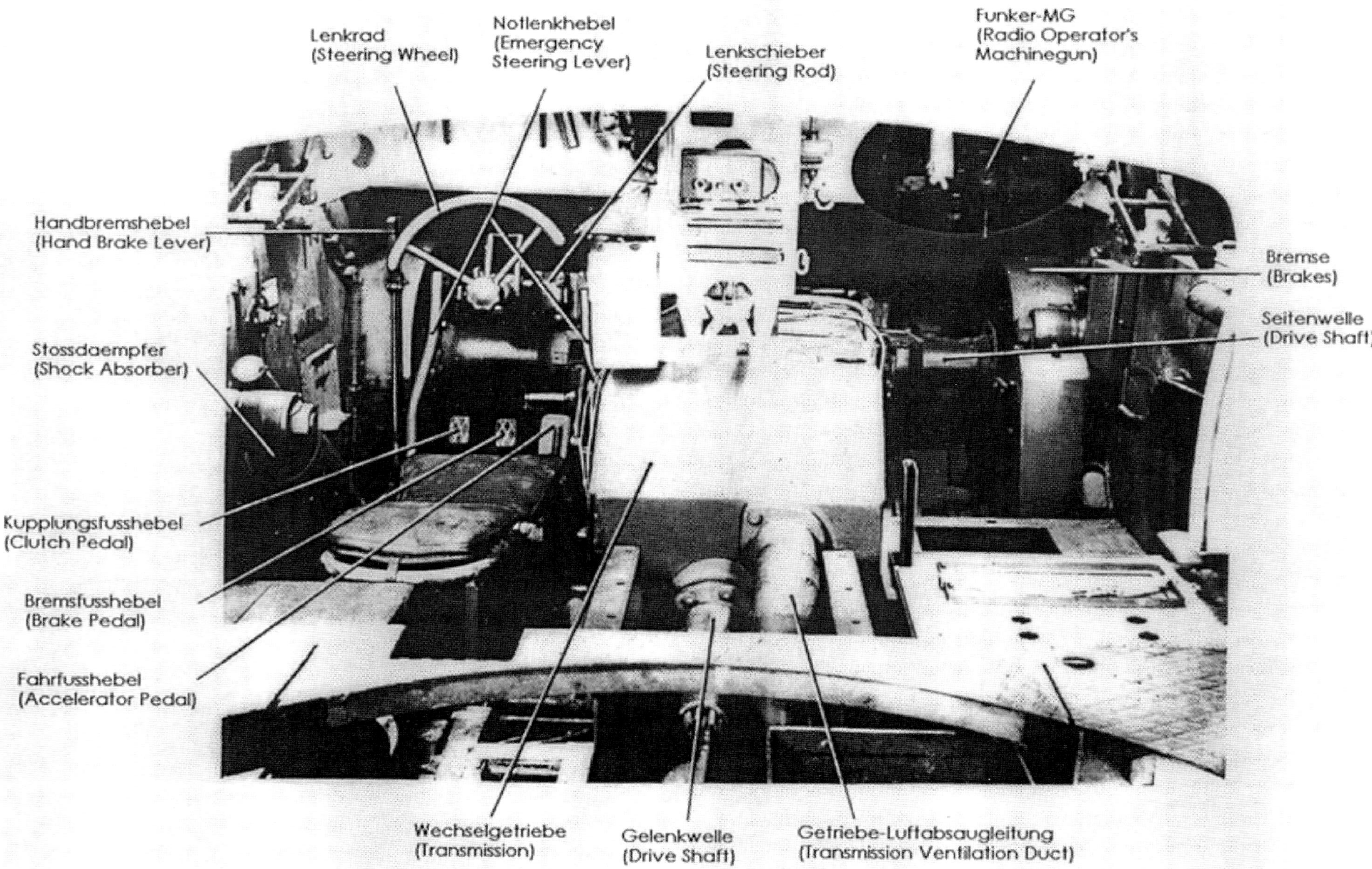

Arrangement of components on the firewall of the Tiger II after several modifications. (Compare with the photograph of the firewall in **Tiger Ausf.B, Fgst.Nr.V2**.)

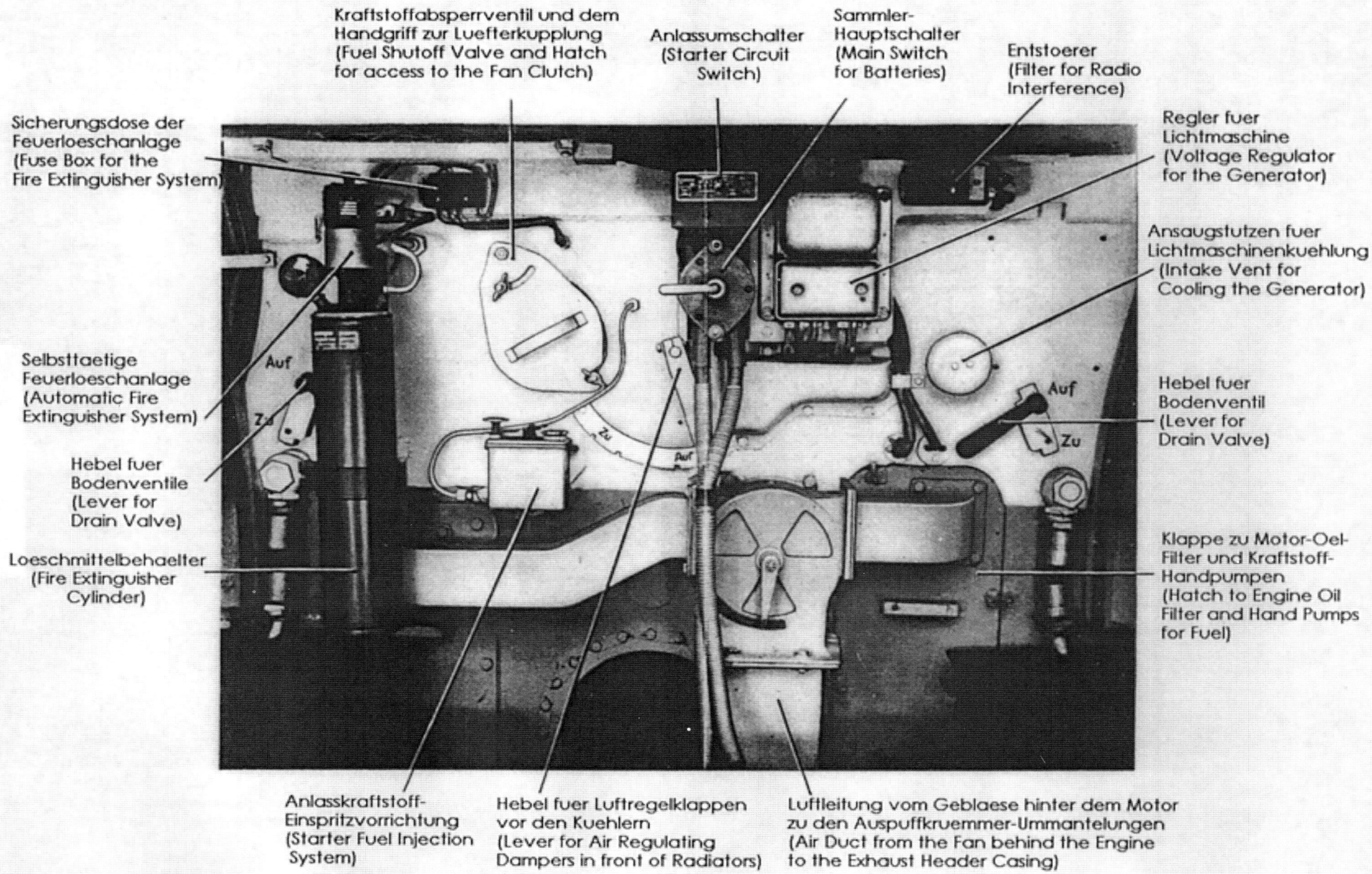

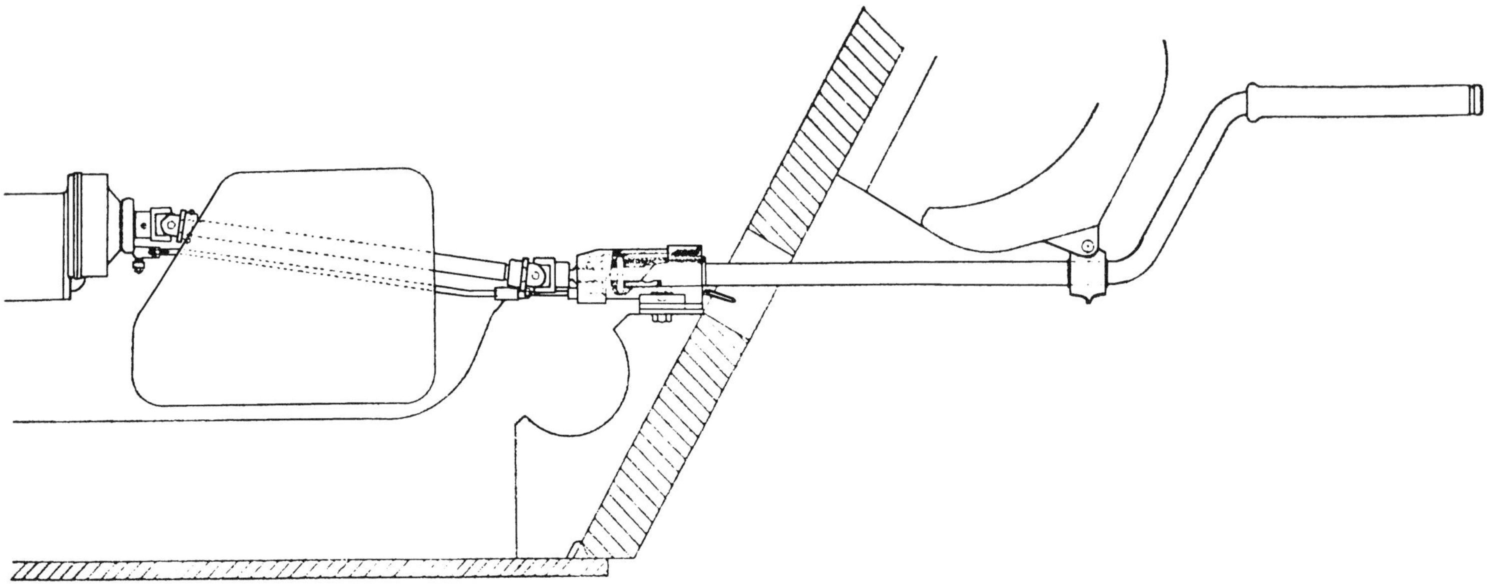

A **Durchdrehanlasser** (hand-cranked starter) was installed for the Maybach **HL 230 P30** engine. To start the engine with the hand crank, the armor cover was removed, the support guide installed, the hand crank installed and engaged with the engine by pulling on the ring.

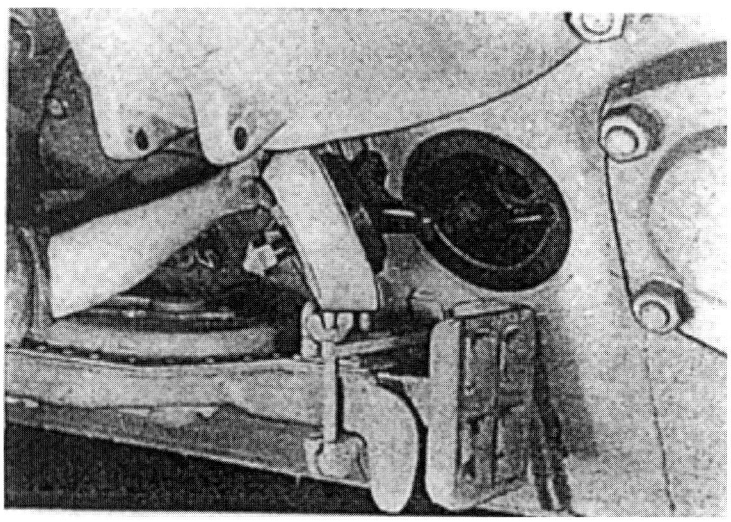

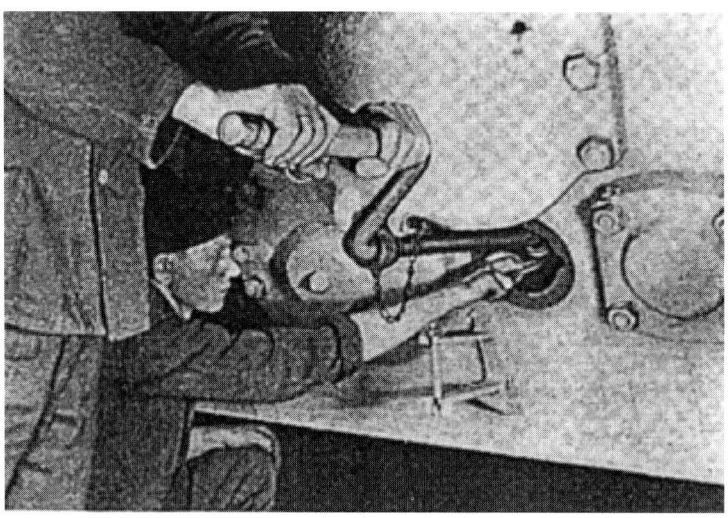

6.1.2 TURM NR. 1 - 50

The following detailed description and drawings of the "Porsche" turret designed by Krupp and mounted on the first 50 Tiger II chassis (**Fgst.Nr.V1-V3** and **280001-280047**) are extracted from the manual **D656/42**, **Panzerkampfwagen Tiger Ausfuehrung B, Turm Nr. 1 - 50 und Panzerbefehlswagen, Geraetbeschreibung und Bedienungsanweisung zum Turm** (equipment description and operating instructions) dated 1 February 1944.

The turret, outfitted with an **8.8 cm Kw.K.43** and a **M.G.34**, was positioned in the middle of the **Panzerkampfwagen**. The gun tube with recoil cylinder and recuperator as well as the **M.G.34** and the gun sight were mounted in the gun cradle. Elevation of the gun cradle was controlled by the hand-operated elevation gear. The elevation arc stretched from -8° to +15°. The frontal weight of the gun was compensated for with a pneumatic counterbalance.

The turret could be traversed either by a hydraulic drive or by hand by the gunner. With the help of the azimuth indicator (12-hour system), the commander could convey traversing orders to the gunner. All openings and joints in the turret could be closed watertight.

The main components of the turret were the:

Turm and Turmanschluss (turret and turret ring)
Wiegenpanzer (gun shield)
Rohrzurrung (internal travel lock)
MG-Lagerung (machinegun mount)
Munitionszufuehrung und Huelsenschleuse (ammunition feed and spent cartridge chute)
Turmzielfernrohr (gun sight)
Prismenspiegellagerung (periscope mount)
Turmschwenkwerk (traverse drive)
Hoehenrichtmaschine (elevation mechanism)
Abfeuervorrichtungen (firing circuit)
Kommandantenkuppel (commander's cupola)
Turmlukendeckel, oberer (loader's hatch)
Turmlukendeckel, hinterer (rear hatch)
MP-Stopfen (pistol ports)
Turmsitze (crew seats)
Munitionslagerung (ammunition racks)
Nahverteidigungswaffe (close defense weapon)
Kompressoranlage (compressor)
Elektrische Einrichtung (electrical equipment)

6.1.2.1 Turm and Turmanschluss (Turret and Turret Ring)

The 360° traversable turret was mounted on the turret ball bearing race. The turret front was curved. The side and rear walls were slanted at 30° from the vertical. The side walls were bent around the turret ring as well as bulged outward around the base of the commander's cupola. Fore and aft, the turret roof was placed at an angle of 78° to the vertical. A circular opening for the commander's cupola, the loader's hatch, and penetrations for the ventilation fan and the **Nahverteidigungswaffe** were located in the middle flat plate on the turret roof. A periscope for the loader was located in the right front angled plate on the turret roof. A hinged port for disposing of spent shell cartridges was located in the rear plate on the turret roof. The embrasure for the machinegun, two openings for the gun sight (**T.Z.F.9b/1**), and the embrasure for the gun penetrated the turret front plate.

The gap between the gun tube and embrasure was shielded by two collars welded to the turret front and by the gun shield. The embrasure for the machinegun and both openings for the gun sight were stepped in order to deflect any bullets or bullet splash.

The rear turret wall was designed to be removed so that the gun could be installed and dismounted. The opening in the rear turret wall was covered by an armor hatch. A pistol port cut into the rear hatch was closed by an armor plug.

A support ring was welded to the bottom of the turret upon which the turret rested on the turret ball bearing race.

Three lifting eyes were welded to the turret roof for removing and mounting the turret.

The device holding the turret on the armored hull was called the **Turmanschluss**. The components of the **Turmanschluss** were the **Zahnkranz** (gear ring), **Turmkugellager** (turret ball bearing race), **Dichtschlauch** (sealing tube), **Tragring mit Abdeckblech** (support ring with cover guard), **Turmzurrung** (turret travel lock), **Zwoelfuhrzeiger** (azimuth indicator - 12 hour), and **Drehbuehne** (turret platform).

The **Zahnkranz** (gear ring) served to turn the turret with the traverse drive. It had 208 teeth. The gear ring was bolted to the armor hull and the outer ball bearing race. Three assembly studs were set into the gear ring which fit into holes drilled in the support ring to make it easier to correctly mount the turret.

The **Turmkugellager** (turret ball bearing race) consisted of two grooved steel rings, in between which ran 70 supporting ball bearings (45 mm diameter) and 70 spacing ball bearings (43 mm diameter). The supporting ball bearings bore the turret weight and took up any side movement. The fixed outer ball bearing ring fitted into and was bolted to the gear ring. The moving inner ball bearing ring fitted into and was bolted to the turret support ring.

The **Dichtschlauch** (sealing tube) was located in the outer ball bearing ring. It served to seal the **Turmanschluss** watertight during under water travel. The sealing tube was filled with air through the air valve located forward on the left, causing the gap between the support ring and the ball bearing race to be sealed. Water seeping into the ball bearing race was drained out through a tube that was sealed by a threaded cap.

The turret was supported on the ball bearing race by the **Tragring** (support ring). A four-piece guard was bolted to the support ring to cover the gear ring. Two mounting plates were located at the front of the support ring for mounting the gun's trunnion mounts. The turret travel lock was on the right, the azimuth indicator on the left, the azimuth indicator drive and the commander's seat on the left rear. In the middle of the support ring, fore and aft, were the traverse drive carriers from which the turret platform was hung. The support for the elevation gear was located on the right side.

The **Turmzurrung** (turret travel lock) was constructed as a geared piece that served to hold the turret in any position desired. The turret travel lock with its geared piece, spindle, and hand wheel was bolted to the support ring. The geared piece was prevented from turning by a positioning bolt. The turret travel lock was operated by turning the hand wheel clockwise, which pushed the geared section into the turret gear ring.

The **Zwoelfuhrzeiger** (azimuth indicator) was mounted to the left front on the support ring. The housing was bolted to a socket and enclosed with a cover. The azimuth indicator needles were moved by a series of internal gears that were driven by the turret gear ring. The azimuth indicator had two dials, one circle divided into 100 and the second circle with an outer ring divided into hundreds of mils (1-64) and an inner ring divided into hours (1-12).

The **Drehbuehne** (turret platform) was made up of the traverse drive carriers and the platform pieces bolted to them. The turret platform was suspended fore and aft by the traverse drive carriers. It supported, as stated by the name, the traverse drive with the hydraulic drive motor. The elevation gear was mounted on the support bolted on the right side. The control for the hydraulic drive motor was mounted on the left front of the turret platform and behind it the compressor. The area underneath the turret platform was accessible through two doors.

6.1.2.2 Wiegenpanzer (Gun Shield)

The gun shield was slid over the gun guide tube and held by two sliding springs and two bolts. Inside the gun shield at the front was a wiper supported by a spring to prevent fouling of the guide tube. The wiper was held by the front armor ring threaded into the gun shield.

The gun embrasure was sealed at the turret front plate. The seal was bolted to the inside of the turret front plate and fastened around the gun cradle by a tightened band.

6.1.2.3 Rohrzurrung (Travel Lock)

The gun was held motionless by an internal travel lock. The travel lock was fastened to the underside of the turret roof in a pivoting mount. To operate the travel lock, the gun was elevated (15°) until the bottom rested on the stop on the elevation gear mount. Then the lower end of the travel lock was positioned in the recess in the gun breech and tightened. When not in use, the travel lock was to be stowed under the turret roof.

6.1.2.4 MG-Lagerung (Machinegun Mount)

The **M.G.34** was mounted on the right side of the gun carriage. The mounting was adjustable in a spring-tensioned housing. The machinegun embrasure could be sealed watertight by installing and tightening a sealing plug in the hole after the machinegun had been dismounted.

6.1.2.5 Munitionszufuehrung und Huelsenschleuse (Ammunition Feed and Spent Cartridge Chute)

The ammunition feed for the machinegun was mounted on an arm of the right gun carriage mount. Two belt sacks, each holding 150 rounds of belted machinegun ammunition, were located in a holder on the turret support ring close to the ammunition feed mechanism. The spent cartridges were directed by a chute into a container located on the right front of the turret platform.

6.1.2.6 Turmzielfernrohr 9b/1 (Binocular Gun Sight)

The **T.Z.F.9b/1** was a jointed, binocular gun sight for direct fire, with an armor plate at the joint to protect the gunner. The sighting angle for the various ranges was changed by adjusting the sighting mark seen in the viewing field. The head of the gun sight could be moved through an arc of +30° to -20°. Each of the two telescopes had an optical length of 814 mm. The magnification was 2.5x with a field of view of 25° (equal to 444 meters wide at a range of 1000 meters) through the 5 mm diameter exit pupil. The focus was adjustable to the gunner's vision.

The reticle in the right-hand telescope consisted of a central inverted V mark flanked by three smaller inverted Vs to the right and left spaced at 4 mil intervals to aid in leading moving targets. Range markers were inscribed in arcs for the various types of ammunition up to 6000 meters. A reticle (with only the inverted V marks) in the left-hand telescope could be rotated into place if the right-hand telescope was unusable. An adjustable light was provided to illuminate the reticles in dim light.

The front of the binocular **T.Z.F.9b/1** was mounted on the slides in the gun sight mount and the rear was suspended from the turret roof by a pivoting arm. The forward gun sight mount was fastened to the gun carriage and pivoted with it. A wiper for cleaning the front protective glass was mounted to the right side of the forward gun mount slides. For underwater travel, both of the openings in the turret front plate for the binocular gun sight could be sealed with plugs.

6.1.2.7 Prismenspiegellagerung (Periscope Mount)

The loader's periscope mount was located in front of the loader's hatch in the turret roof. The periscope in its bakelite housing was pressed against the rubber seal by tightening the wing nuts on each side of the holder. An armor hood was welded onto the turret roof to protect the periscope from above.

6.1.2.8 Turmschwenkwerk (Traverse Drive)

The turret was turned by operating the traverse drive. The traverse drive was powered by the vehicle engine, but it could also be turned by hand.

A power takeoff from the transmission drove the hydraulic motor for the traverse drive. Powered traverse was controlled by a foot pedal mounted on the turret platform. A hand wheel was used to select high or low range, and a hand-operated clutch engaged and disengaged the drive.

An auxiliary drive, mounted on the right side of the turret platform, could be used by the loader to assist the gunner in traversing the turret by hand.

One turn on the gunner's hand wheel or loader's auxiliary hand wheel resulted in traversing the turret 0.5° or about 9 mils.

6.1.2.9 Hoehenrichtmaschine (Elevation Mechanism)

The elevation mechanism was a spindle elevating gear. The gunner's handwheel turned beveled gears driving a threaded shaft which raised or lowered the gun carriage. One turn of the hand wheel changed the gun's elevation by about 0.65° or 11.5 mils. The gun trigger and the firing circuit switch were located close to the elevating hand wheel.

6.1.2.10 Abfeuervorrichtungen (Firing Circuit)

A firing lever for the main gun was mounted on the elevation handwheel. The contacts in the electrical firing circuit were closed when the firing lever was pulled. When the lever was released, a spring returned the lever to its rest position and the firing contacts were opened.

The machinegun was fired by stepping on a foot pedal mounted on the turret platform. The foot pedal was connected by a cable to a mechanical firing lever for the machinegun. A spring returned the foot pedal to the rest position.

6.1.2.11 Kommandantenkuppel (Commander's Cupola)

The commander's cupola was located in the middle on the far left side of the turret roof. It served as the entrance hatch and provided sheltered observation for the commander. The commander's cupola consisted mainly of the armor ring, seven periscopes, hatch lid with pivoting arm, and azimuth indicator ring.

The armor ring was set into a circular cutout in the turret roof and fastened with bolts. Seven evenly spaced openings were cut into the armor ring for mounting seven periscopes in their bakelite housing. The bakelite housings and their flanges were pressed against the rubber seals by tightening two wing nuts. An armor hood was welded over each periscope to protect it against damage from above.

The pivoting arm with the hatch cover was released by turning a handwheel that operated a threaded spindle. The operating lever was then used to raise the hatch cover and swing it back to the rear and left.

The azimuth indicator ring was carried by six rollers on the inside of the armor ring. It was marked from 1 to 12 o'clock and driven by a shaft connected to a drive gear on the turret ring.

6.1.2.12 Turmlukendeckel, oberer (Loader's Hatch)

The hatch in the turret roof served for entering and leaving the turret. The hatch cover lay in a cutout recess in the turret roof and was held by two hinges. There was a handhold on both the outside and inside. To close the hatch watertight, four locking bars were pushed outward by a central ring, and either the handwheel from the inside or a square key from the outside were used to tighten the central threaded spindle to pull the hatch cover down onto the rubber gasket.

6.1.2.13 Turmlukendeckel, hinterer (Rear Hatch)

The hole in the rear wall of the turret facilitated installation and removal of the main gun. This hole was covered by a bolted-on frame and a hinged hatch cover. Torsion bars were installed into the hinge pins to make it easier to open and close the heavy hatch cover. A cable with a handle was provided to pull the hatch closed from the inside. Two hand-operated bolts were used to seal the hatch cover against the watertight rubber gasket.

A pistol port cut into the hatch cover was closed with an armor plug. A sheet metal slide guard on the inside of the hatch was mounted on a hinge at the bottom and held in place by a latch. This guard had to be unlatched to gain access to the pistol port.

6.1.2.14 MP-Stopfen (Pistol Port)

The pistol port in the rear turret hatch cover was closed with an armor plug. The armor plug was held in position by a collar that was rotated to the side to release the armor plug. Two chains were attached to the armor plug. The shorter chain held the armor plug after it was pushed out. The longer chain was used to pull the armor plug back into place.

6.1.2.15 Turmsitze (Crew Seats)

Kommandantensitz (Commander's Seat) - A hinged saddle seat with back cushion and footrest for the commander was mounted on the left side of the turret.

Richtschuetzensitz (Gunner's Seat) - A saddle seat with back cushion for the gunner was mounted on the left side of the turret platform. The saddle seat was mounted on a support that also carried the clamp mount for the hand traverse housing.

Ladeschuetzensitz (Loader's Seat) - A saddle seat for the loader was located on the right side of the turret platform.

6.1.2.16 Munitionslagerung (Ammunition Racks)

Sixteen **8.8 cm Patronen** (cartridges) were stored in the turret rear, eight on the left and eight on the right side.

6.1.2.17 Nahverteidigungswaffe (Close Defense Weapon)

The **Nahverteidigungswaffe** was used to fire the **Sprenggranatpatrone 326 Lp** (explosive shells), **Schnell-nebelkerzen 39** (quick smoke candles), **Rauchsichtzeichen orange 160** (orange smoke signals), and **Leuchtgeschossen R** (flare signals) with the **Zundschraube C 43 St** (propellant charge). The **Nahverteidigungswaffe**, aimed at a fixed angle of about 50°, could be traversed 360°. An armor plug could be inserted into the bore to seal it when not in use.

The **Sprenggranatpatrone 326 Lp** with a 1 second delay fuze was propelled out to 7 to 10 meters and exploded about 0.5 to 2 meters above the ground. Individual fragments flew in a circle for up to 100 meters. All turret hatches and openings were to be closed when this round was fired.

6.1.2.18 Kompressoranlage (Compressor)

The compressor delivered air for blowing the fumes out of the gun tube after firing. It was mounted on the turret platform and driven by belted pulleys by the turret hydraulic traverse drive. The compressor pump was used to fill the air tank (located toward the rear of the turret platform) which was connected by tubing to the gun breech.

6.1.2.19 Elektrische Einrichtung (Electric Equipment)

The firing circuit was powered by the 12 volt vehicle battery. If the circuit failed, an emergency inductive charge firing device was used.

Turret lighting was provided by three lights mounted under the roof. The amount of light could be adjusted by rotating covers which turned off the light when they were completely closed. Lights were also provided for the gun sight reticles and the azimuth indicator. Two receptacles were available to plug in hand-held lamps.

A ventilator fan with a flow rate of 12 m^3/min was mounted in the turret roof. It served to suck out any fumes left in the turret after the guns were fired.

6.1.3 EQUIPMENT

The following list of equipment was extracted from the **Vorlaeufige Ausruestung fuer Aufbau des Pz.Kpfw.Tiger, Ausfuehrung B**, current as of October 1943:

Fliegerbeschussgeraet fuer M.G.34 (Anti-Aircraft MG Mount)
Regenschutz fuer Kommandanten-Kuppel (Rain Guard)
Turmsehrohr T.S.R.1 (Observation Periscope)
3 Sprengpatronen Z85 (Charges to Demolish the Tiger)
Fuehrerwimpel (Hakenkreuz) (Commander's Pennant)

Ausfallflagge (gelb-schwarz) (Out-of-Action Flag)
3 M.G.34 mit Panzermantle (Machineguns with Armor Sleeve)
(1 in Turret, 1 in Hull, and 1 in Anti-Aircraft Mount)
32 Gurtsack (Bags for Machinegun Ammunition Belts)
(22 in **Befehlswagen**)
1 M.P.40 (Machinepistol)
1 Leuchtpistole (Signal Pistol)
24 Leuchtpatronen (Signal Flares)
1 Kurskreisel (Gyroscopic Compass)
1 Nahverteidigungswaffe (Close Defense Weapon)
12 Schnellnebelkerze 39 (Quick Smoke Candle)
10 Fliegersichtzeichen Orange 350 (Orange Smoke Signal)
20 Sprenggranatpatronen 326 Lp. (Explosive Shells)

6.1.4 COMMUNICATIONS

When assigned to company headquarters or to a platoon leader, the **Panzerkampfwagen Tiger (8.8 cm Kw.K.43 L/71) (Sd.Kfz.182)** was outfitted with two radio sets, a **FuG 5** (10 watt transmitter with ultra short wave length receiver, operated in the frequency band 27.2 to 33.4 MHz) and a **FuG 2** (ultra short wave length receiver, operating in the same frequency band as the **FuG 5**). The **FuG 5** had a usable range of 4 to 6 kilometers, highly dependent on terrain and atmospheric conditions. The other nine **Panzerkampfwagen Tiger (8.8 cm Kw.K.43 L/71) (Sd.Kfz.182)** in each company of fourteen were outfitted with a single **FuG 5** radio set. A **Bordsprechanlage** (intercom system) was to be installed in all **Panzerkampfwagen Tiger (8.8 cm Kw.K.43 L/71) (Sd.Kfz.182)**. Only four of the crew members had speaker headsets and a microphone. As shown in the wiring diagram for the turret, provisions had not been made for a speaker headset or microphone for the loader. The commander could also use voice, signal pistol, flags, orange smoke signals, and directional fire for communicating with others outside the Tiger II.

6.1.5 PANZERBEFEHLSWAGEN TIGER AUSF.B

The **Panzerbefehlswagen** (command tank) version of the **Tiger Ausf.B** was created simply by installing additional radio sets (along with the necessary transformers, cushioned mounts, wiring, junction boxes, aerials, **GG400** auxiliary generator set and other equipment) and reducing the ammunition stowage to provide the needed space. Compared to a **Pz.Kpfw.Tiger Ausf.B** with 80 rounds of main gun ammunition and 32 bags each containing 150 rounds of machinegun ammunition, the **Pz.Bef.Wg.Tiger Ausf.B** carried only 63 rounds for the main gun (32 **Pzgr.** (AP) and 31 **Sprgr.** (HE)) and 22 bags for the machineguns.

The **Panzerbefehlswagen Tiger Ausf.B** was outfitted in two versions, the **Sd.Kfz.267** for longer range communication with higher headquarters and the **Sd.Kfz.268** for communication with supporting aircraft. The **Sd.Kfz.267** was outfitted with a **FuG 8** (30 watt transmitter with medium wave length receiver, operated in the frequency band 0.83 to 3 MHz) and a **FuG 5** (10 watt transmitter with ultra short wave length receiver, operated in the frequency band 27.2 to 33.4 MHz). A **Sternantenne D** (star aerial) for the **FuG 8** was mounted on an **Antennenfuss beweglich Nr.1** (flexible antenna base) with an insulator. The white porcelain insulator was protected by an armor cylinder welded to the rear deck behind the engine compartment vent. An **Antennenstab 2 m** (2-meter rod aerial) for the **FuG 5** was mounted behind the loader's hatch on the turret roof.

The **Sd.Kfz.268** was outfitted with a **FuG 7** (20 watt transmitter and ultra short wave length receiver, operated in the frequency band 42.1 to 47.8 MHz) and a **FuG 5**. A 1.4-meter **Antennenstab 1.4 m** (1.4-meter rod aerial) was used with the **FuG 7** radio set.

Section 23 of D656/42, the manual on **Turm Nr.1-50** dated 1 February 1944, states: *Every tenth **Tiger Ausf.B** is outfitted as a **Panzerbefehlswagen***. However, this did not occur, as confirmed by a note in the **Wa Pruef Technische Daten** sheet G331 that starting with the 51st produced, the **Tiger Ausf.B** was also completed as **Pz.Bef.Wg.Tiger Ausf.B**. D656/42, the manual for the **Serienturm** dated 1 July 1944, doesn't contain any details on radio equipment or its location but simply states that D9023/1, **Funk und Bordsprechanlage im Pz.Kpfw.Tiger als Pz.Bef.Wg.** (created for the Tiger I), was to be used as guidance. A production record from Henschel for September 1944 reveals that every 20th **Tiger Ausf.B (Fgst.Nr.280200, 280220, 280240, 280260**, etc.) was being outfitted as a **Panzerbefehlswagen**.

This first **Versuchsturm** (trial turret), assembled by Krupp for **Wa Pruef** testing and firing trials, already had the **Schauloch** (view port) welded shut but still had a usable **MP-Stopfen** (pistol port). (TTM)

PAGES 43-45: Details of **Turm Nr.150110** mounted on **Tiger II** (**Fgst.Nr.V2**, completed by Henschel in January 1944) now on display at The Tank Museum in Bovington, England (TTM)

The **MP-Stopfen** (pistol port) welded shut on the right side of the turret. (Disregard the cross, tactical number, and camouflage pattern.)

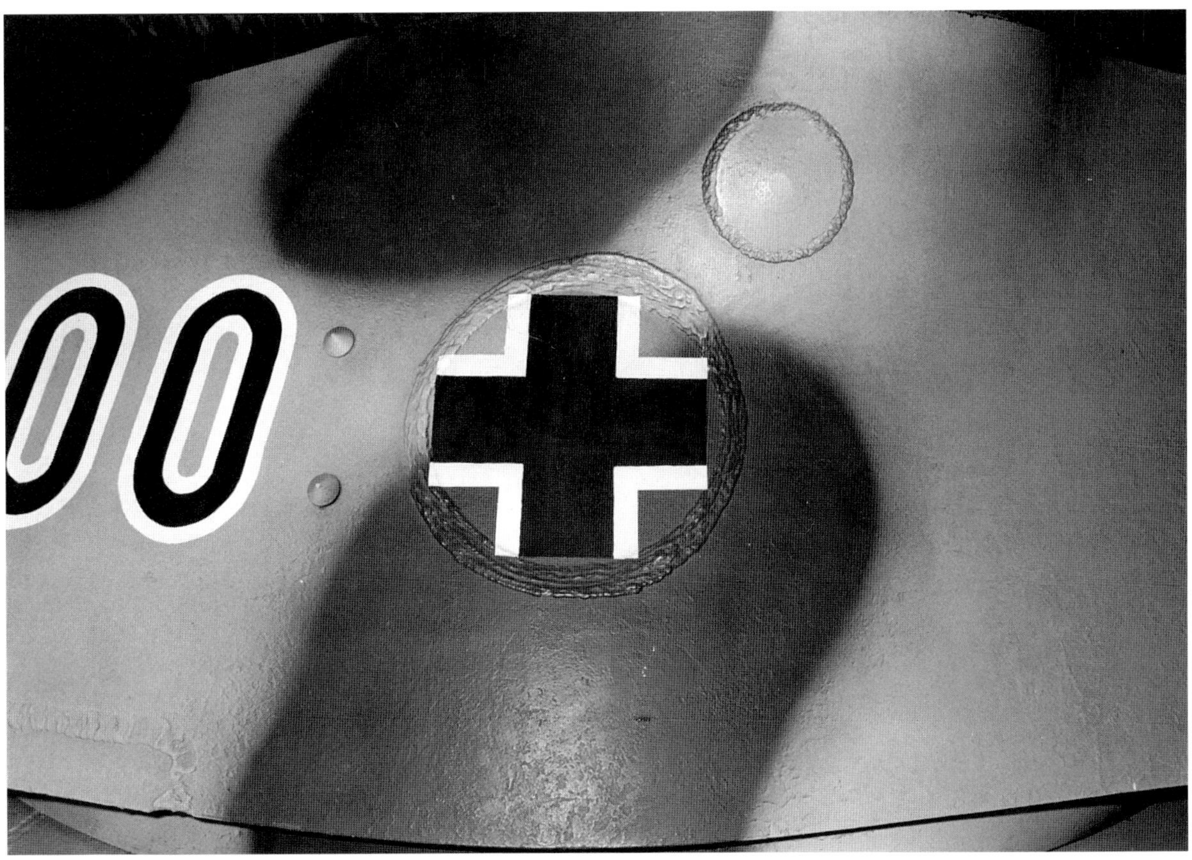

The **MP-Stopfen** (pistol port) and **Schauloch** (view port) welded shut on the left side of the turret.

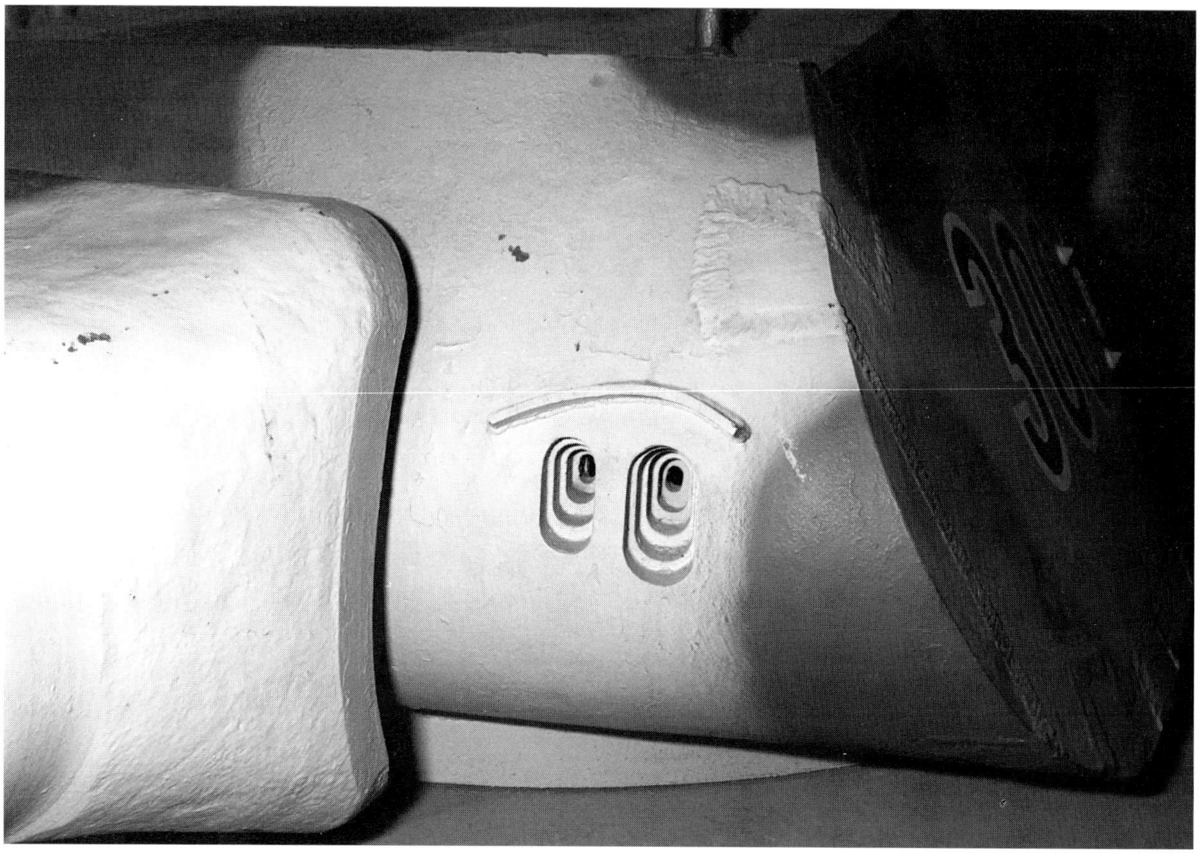

The stepped apertures for the binocular gun sight cut into the 100 mm thick front plate, protected by a rain channel.

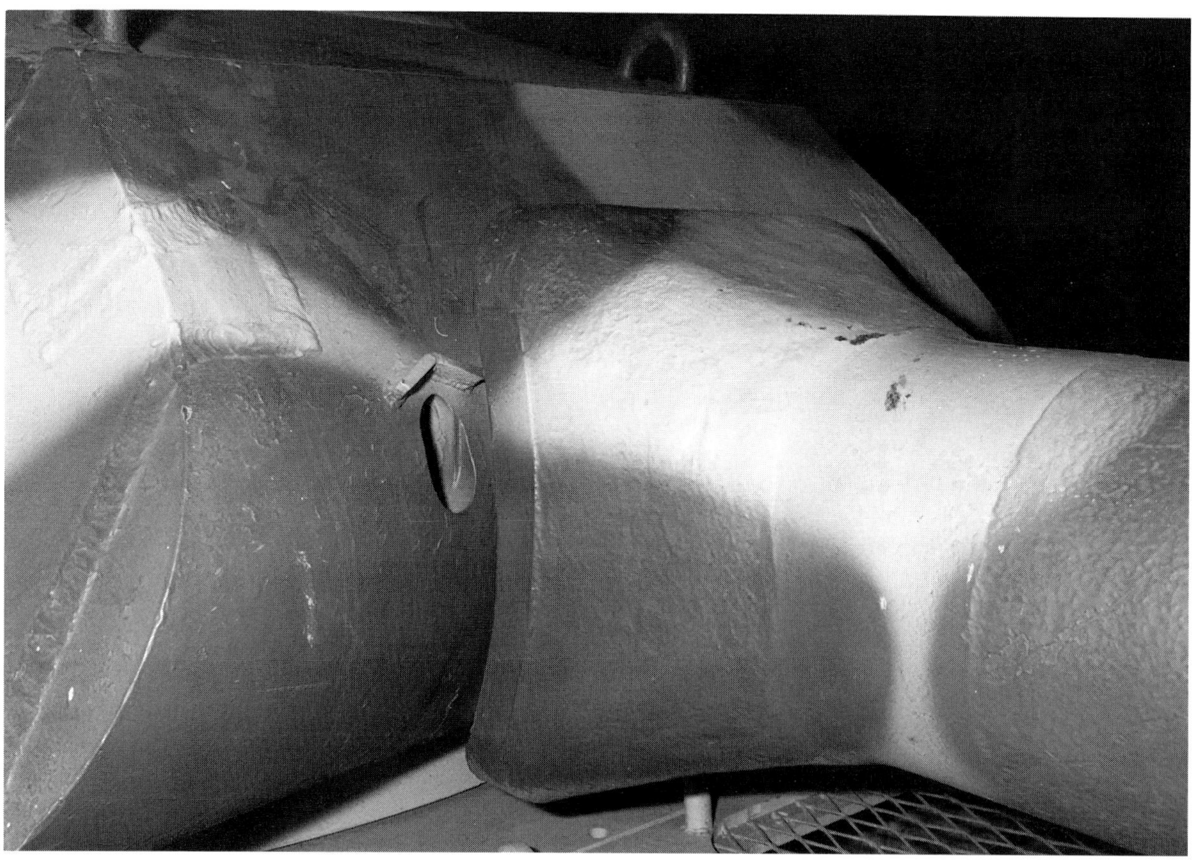

The stepped aperture cut into the 100 mm thick front plate for firing the turret-mounted **M.G.34**. (Disregard the hull roof. The maintenance hatch cover has been installed backwards.)

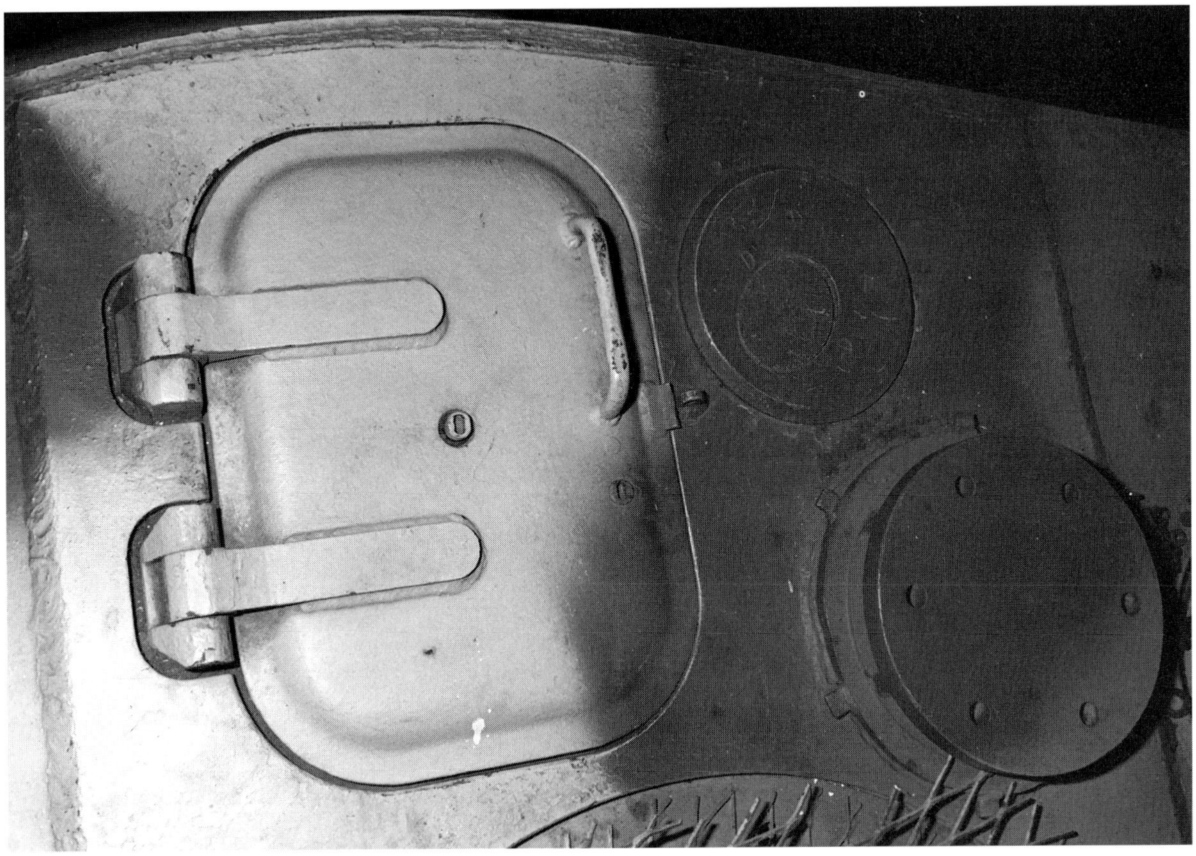

The closed loader's hatch, **Nahverteidigungswaffe** (close defense weapon), and armor guard over the turret vent fan. (The commander's cupola has been removed and the hole covered by metal screening.)

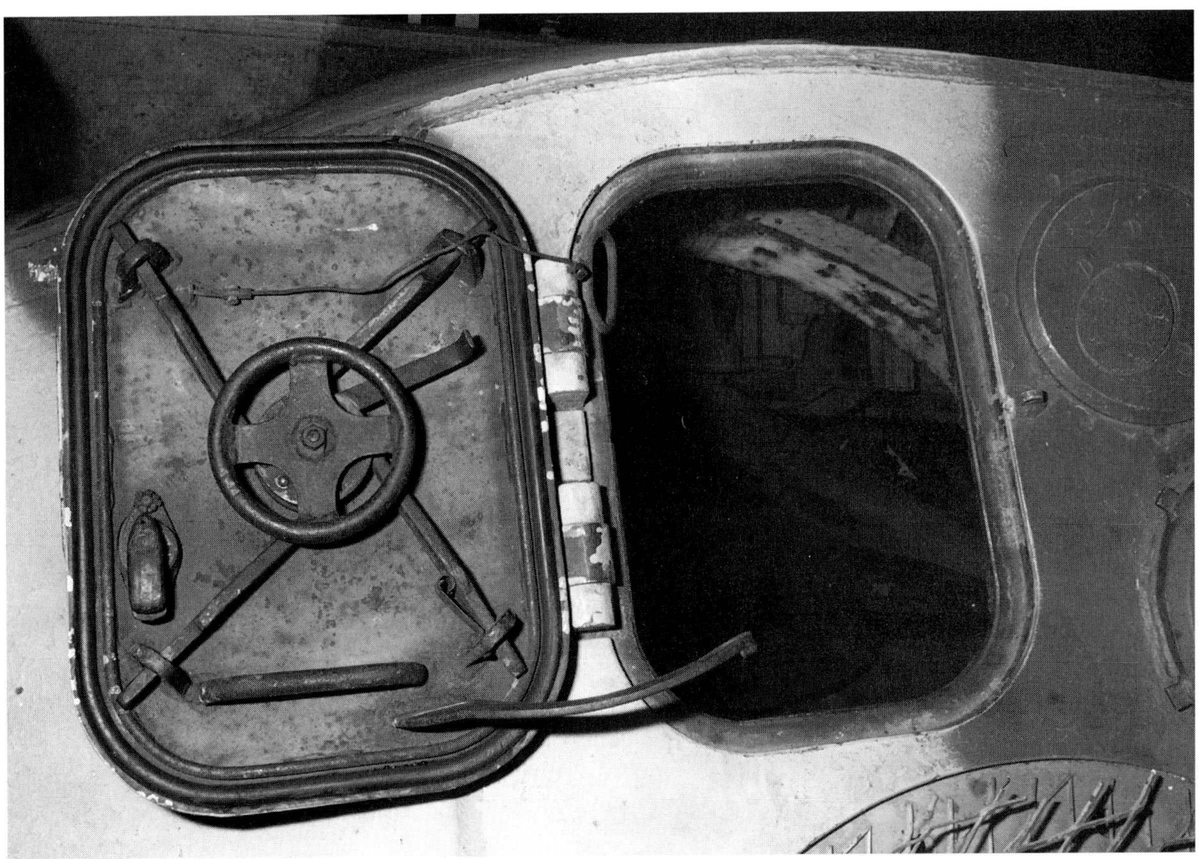

The open loader's hatch with the cable to pull the hatch lid closed, four bars with the handwheel to seal the hatch watertight, two position latch, and curved arm that pushed against a counterbalance spring mounted under the turret roof.

The following are extremely rare photographs of the interior of a **Tiger Ausf.B** with **Turm Nr.1-50**. These photographs were taken inside **Tiger Ausf.B** (**Fgst.Nr.V2**, **Turm Nr.150110**) after it was captured at Haustenbeck. It is now on display at The Tank Museum in Bovington, England, unfortunately stripped of most of its internal fittings.

LEFT: Looking down through the hatch in the commander's cupola. The commander's saddle-shaped seat and padded backrest and the padded backrest for the gunner's seat with the footrest for the commander. RIGHT: Inside the turret looking forward. The binocular **T.Z.F.9b/1** with its headrest is supported from the turret roof by a wishbone-shaped pivoting support. The belted ammunition, fed from a bag mounted on the right side, was guided across the top and fed into the left side of the coaxially mounted **M.G.34**.

LEFT: The eight 8.8 cm rounds stowed in the right rear of the turret were each held in place by two quick-disconnect bands.
RIGHT: The open turret rear hatch with the sheet metal cover, hinges with torsion bar counterbalance, and cable for pulling the hatch cover shut.

8.8 cm rounds stowed in the pannier on the right side. The top rack was inaccessible due to a bar mount used for stowing machinegun ammunition bags.

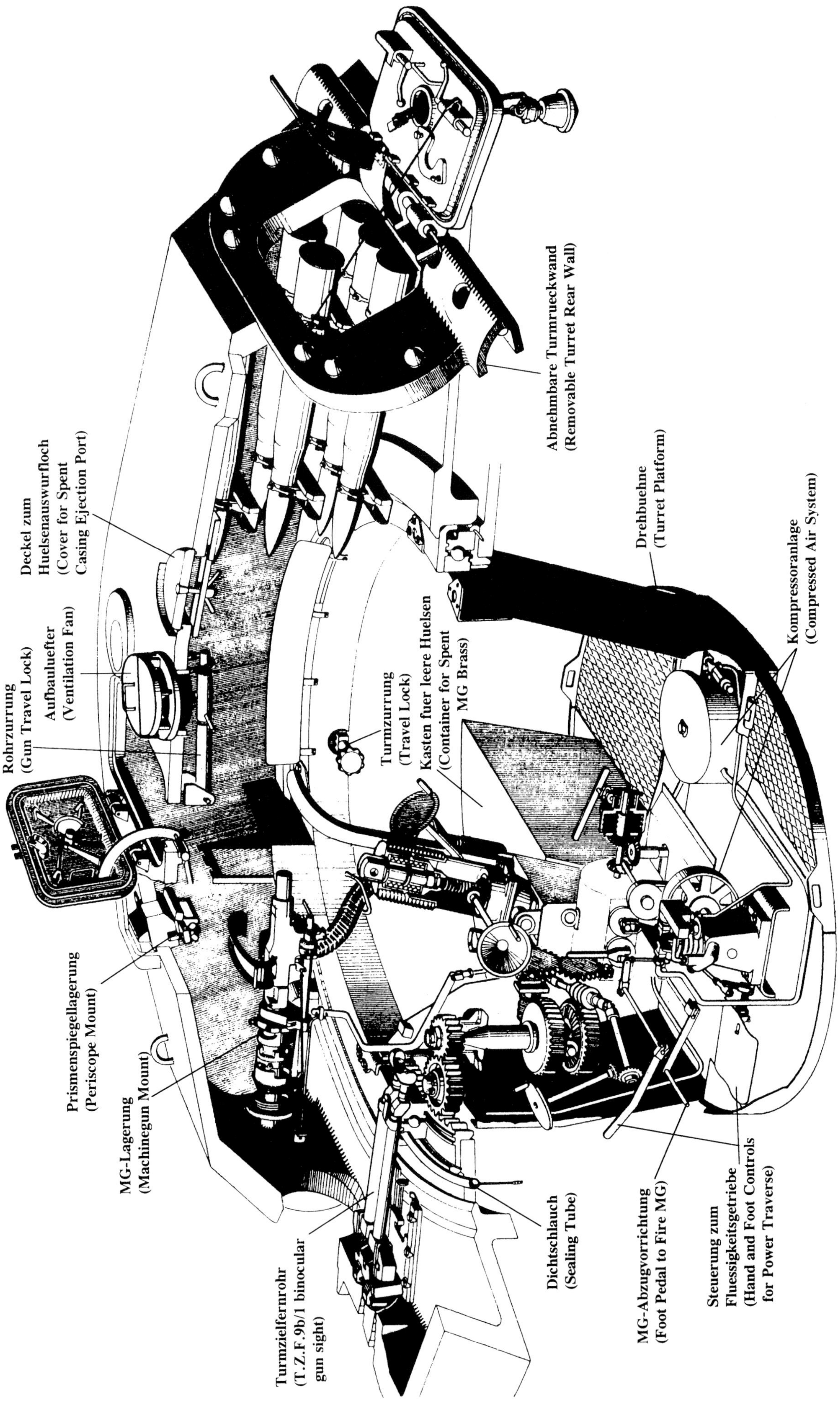

Arrangement of components inside **Turm Nr.1-50** (Not to Scale)

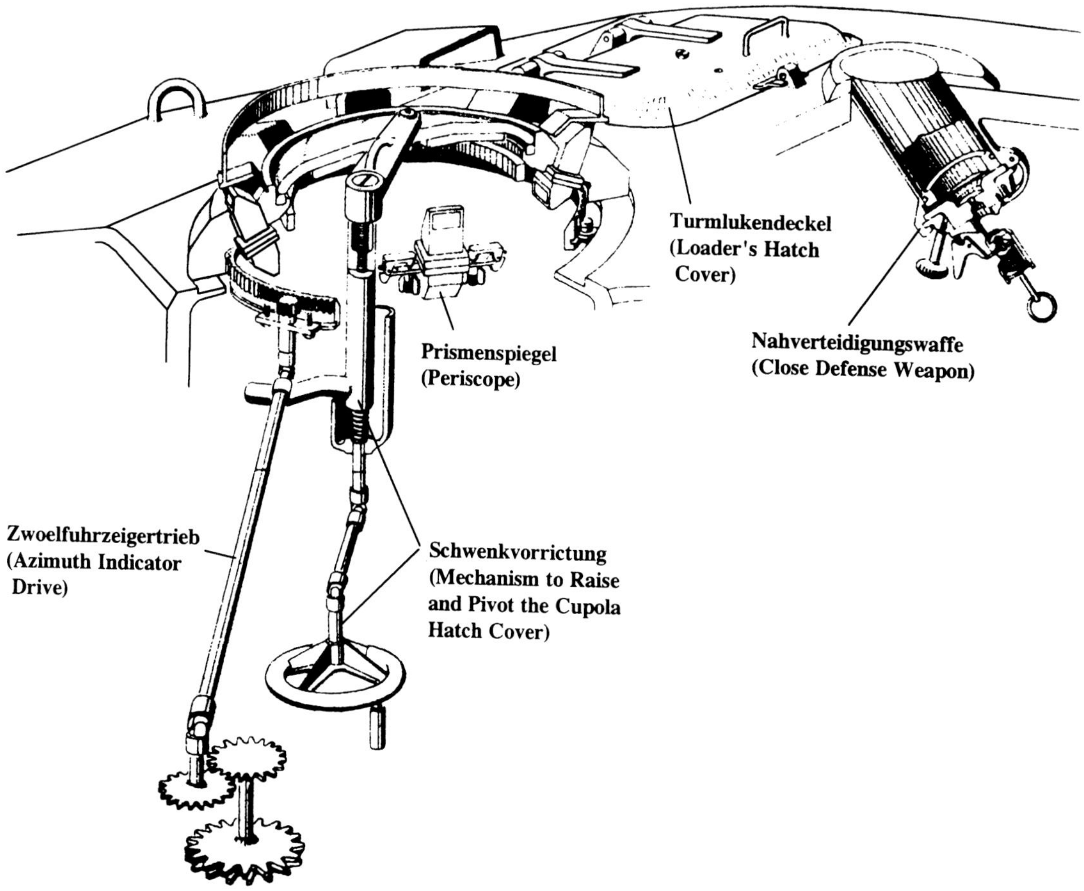

Arrangement of components on the roof of **Turm Nr.1-50** (Not to Scale)

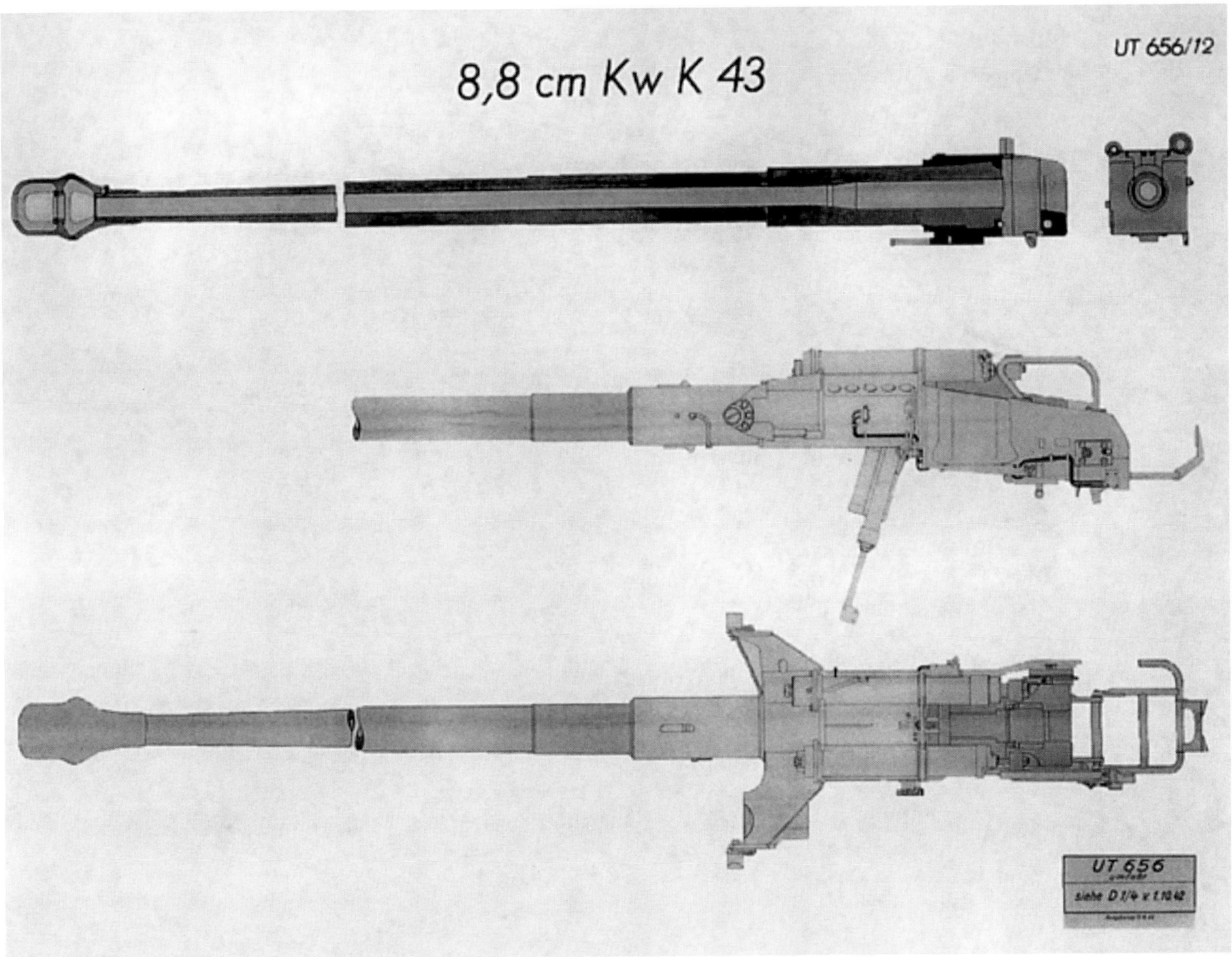

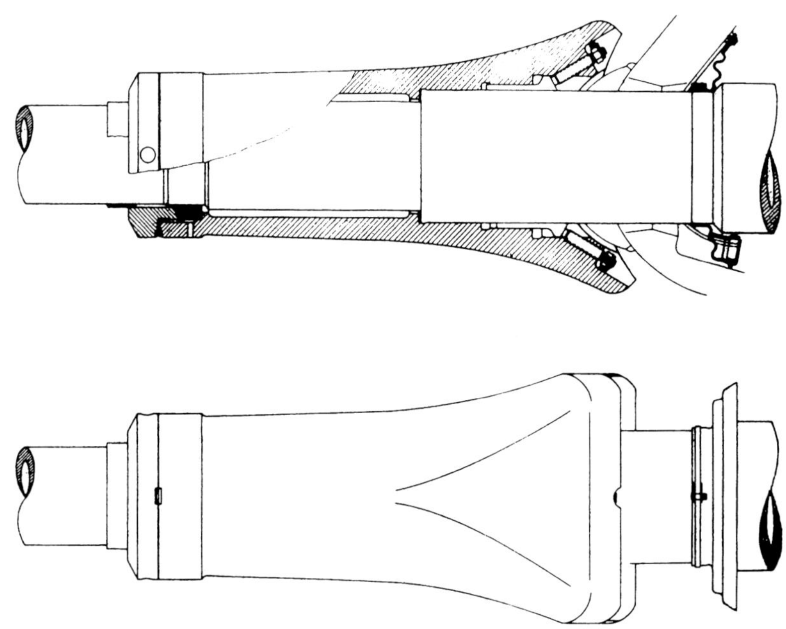

Wiegenpanzer (021 B 48014) gun shield. The sectional views show the way it was mounted and the seal to prevent accumulation of dirt inside the slide.

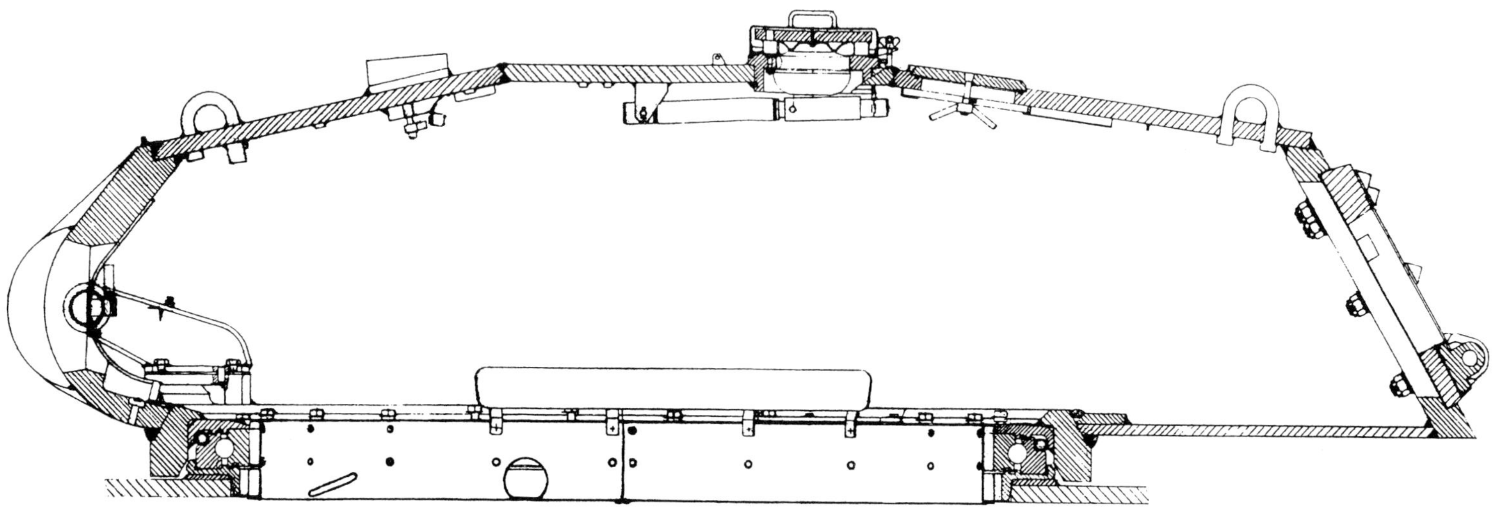

Turmgehaeuse (021 B 48011) turret armor body for **Turm Nr.1-50** with the interior travel lock mounted under the turret roof. The turret front was curved to fit around the mount for the gun trunnions.

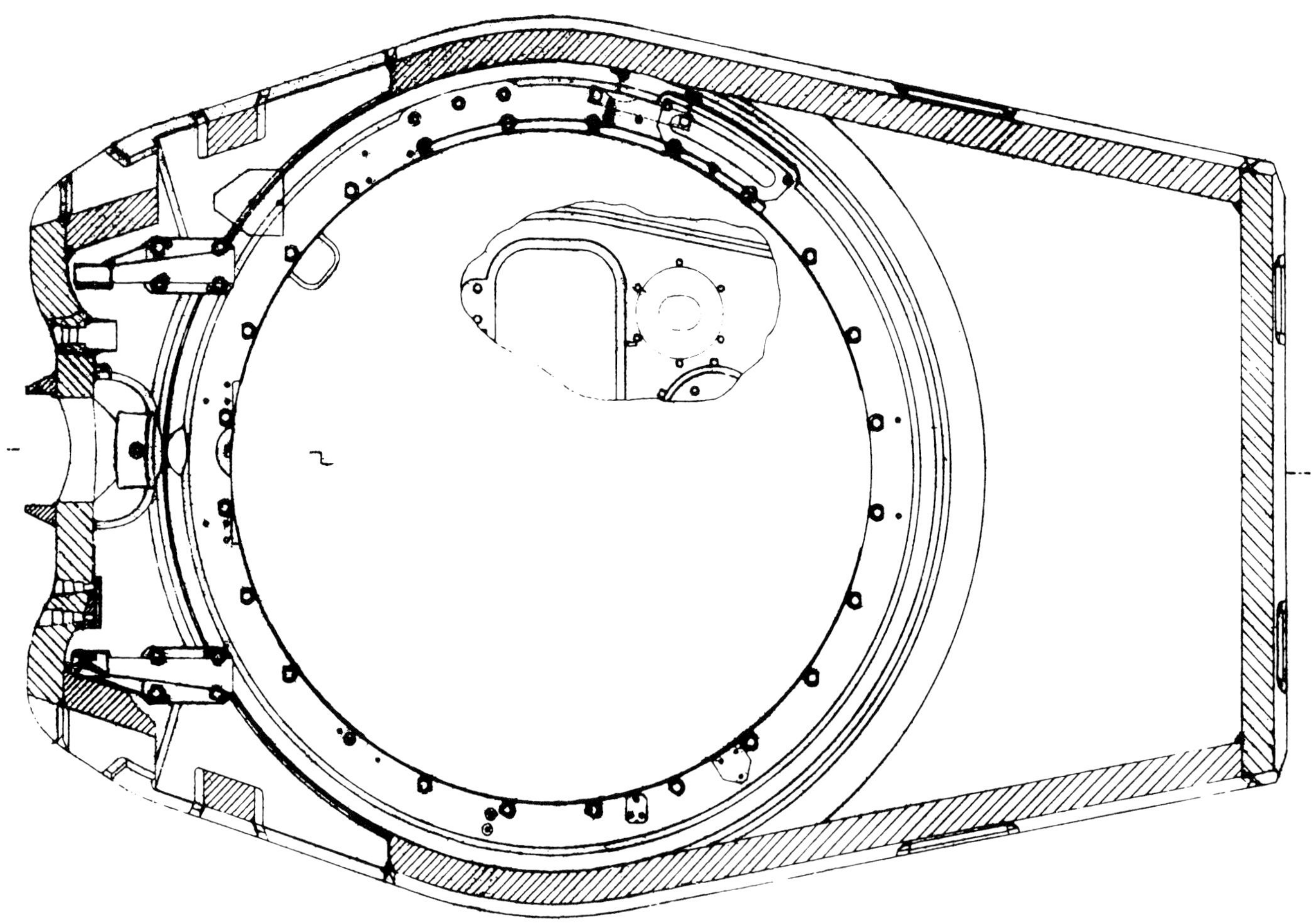

Turmgehaeuse (021 B 48011) turret armor body for **Turm Nr.1-50** with marks showing that the centerline of the gun was offset 30 mm to the right of the turret centerline.

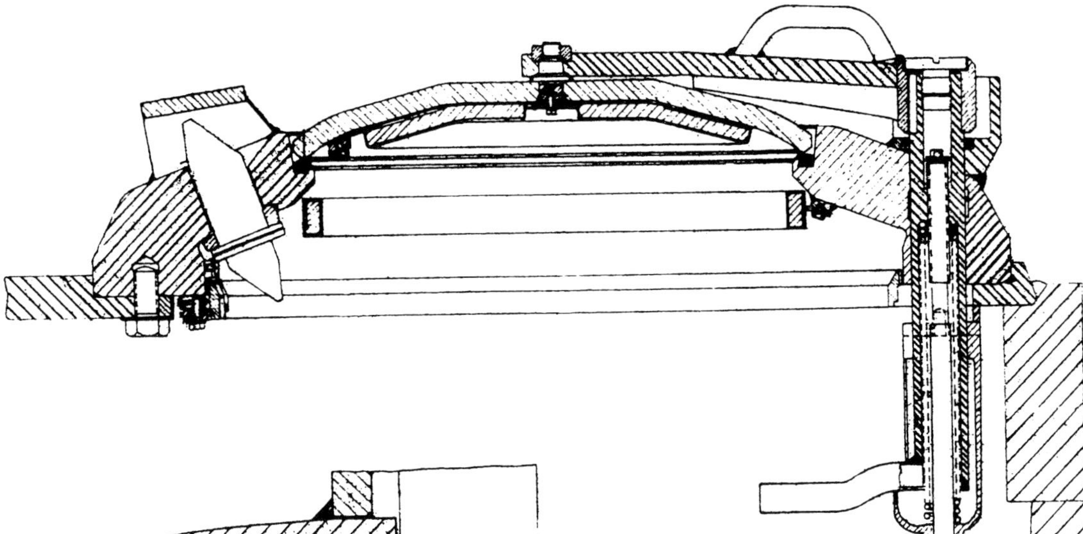

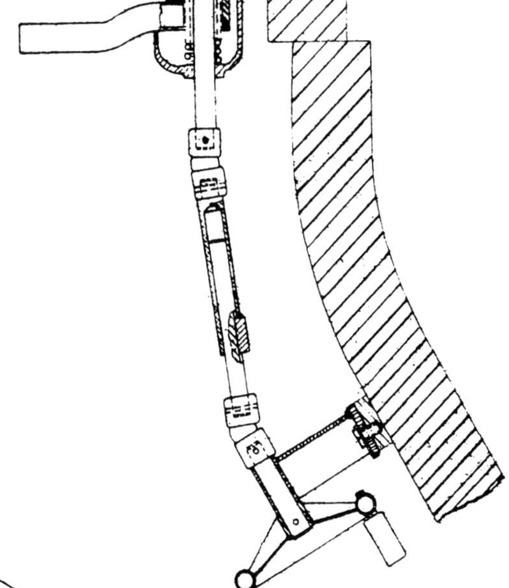

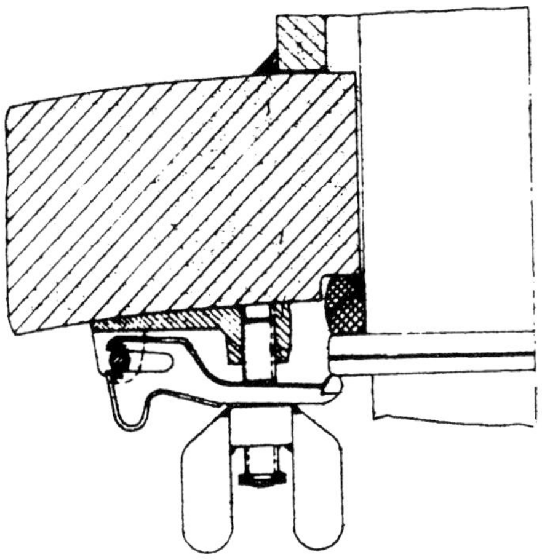

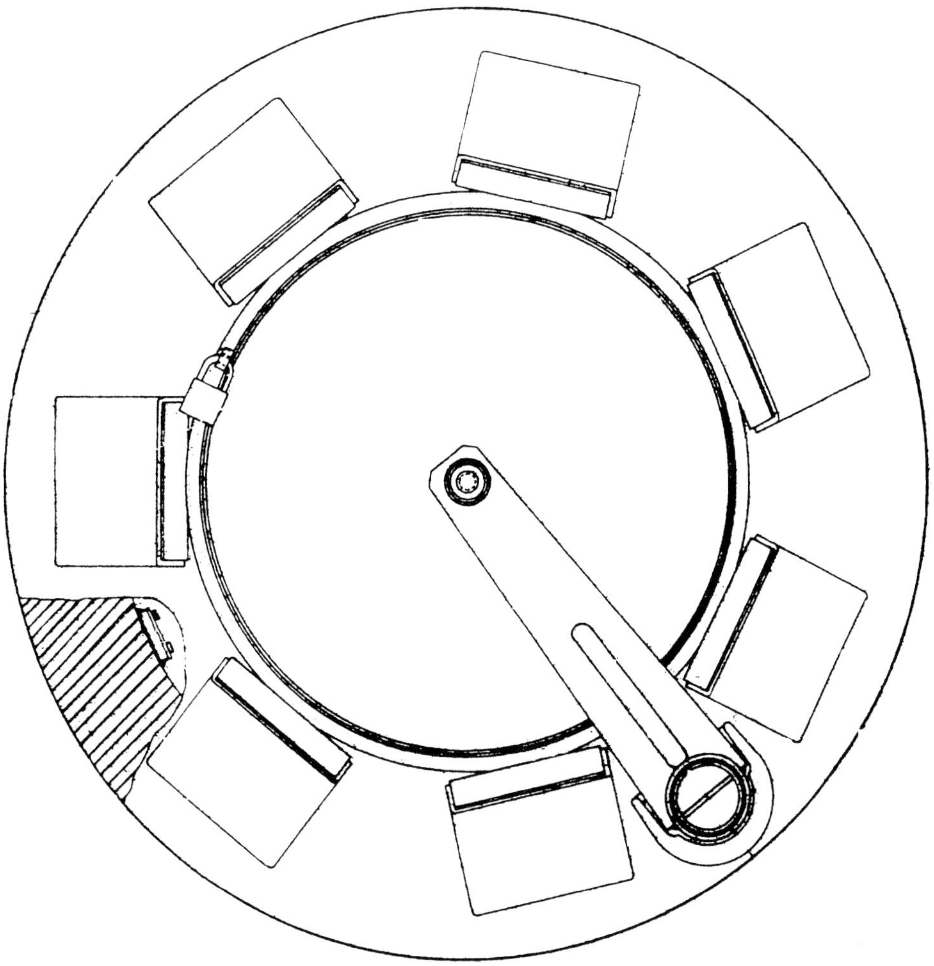

Kommandantenkuppel (021 B 48013) commander's cupola with seven periscopes. To open the hatch, the commander first turned on the handwheel to lift the cover and then pivoted the cover toward the rear using the horizontal handle. The two-piece periscopes were held in place by wing nuts.

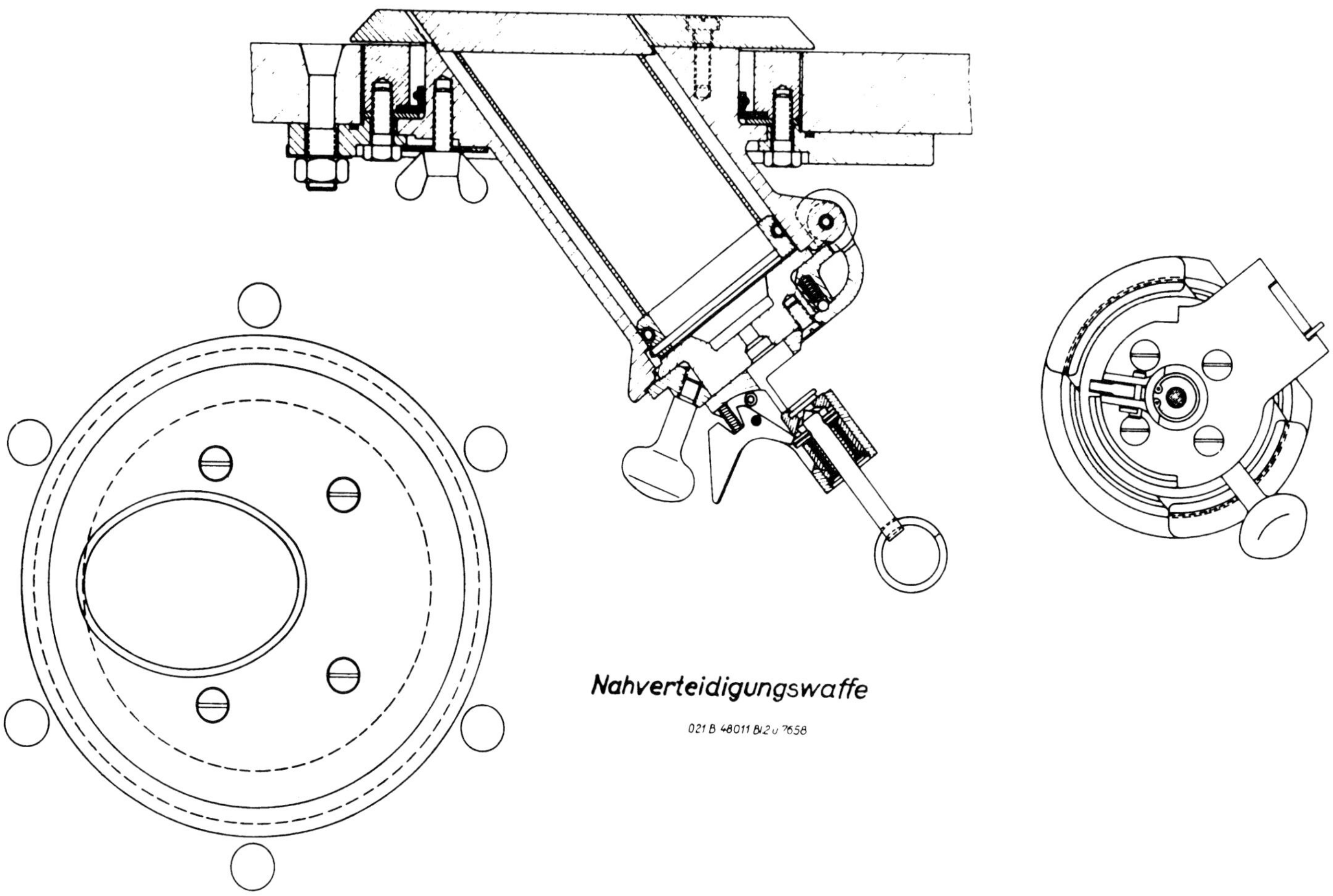

The top, side, and bottom of the breach of the **Nahverteidigungswaffe (021 B 7658)** close defense weapon showing the spring-loaded firing pin in its cocked position, held back by the trigger.

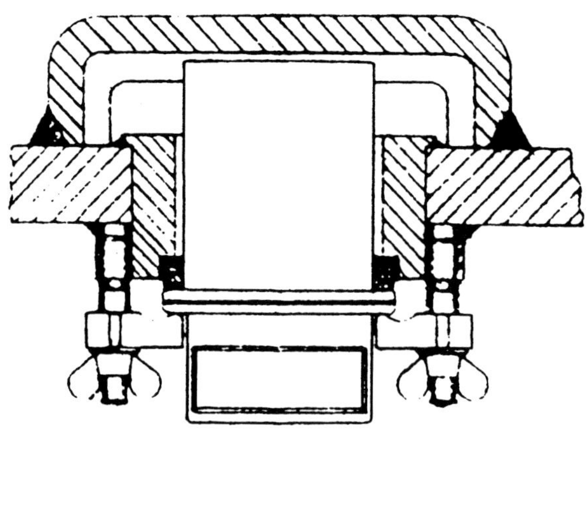

The **Prismenspiegel** (loader's periscope) mounted in the turret roof was protected on the top and sides by an armor cover. The two-piece periscope was held in place by wing nuts.

OPPOSITE BELOW: Sixteen rounds for the main gun were stored in the **Geschuetzmunitionslagerung (021 A 48036)** at the rear of the turret. A sheet-metal shield protected the ammunition from fragments spalling off the inside of the turret wall when hit by AP projectiles. Each round was held in place by two straps with quick release clamps.

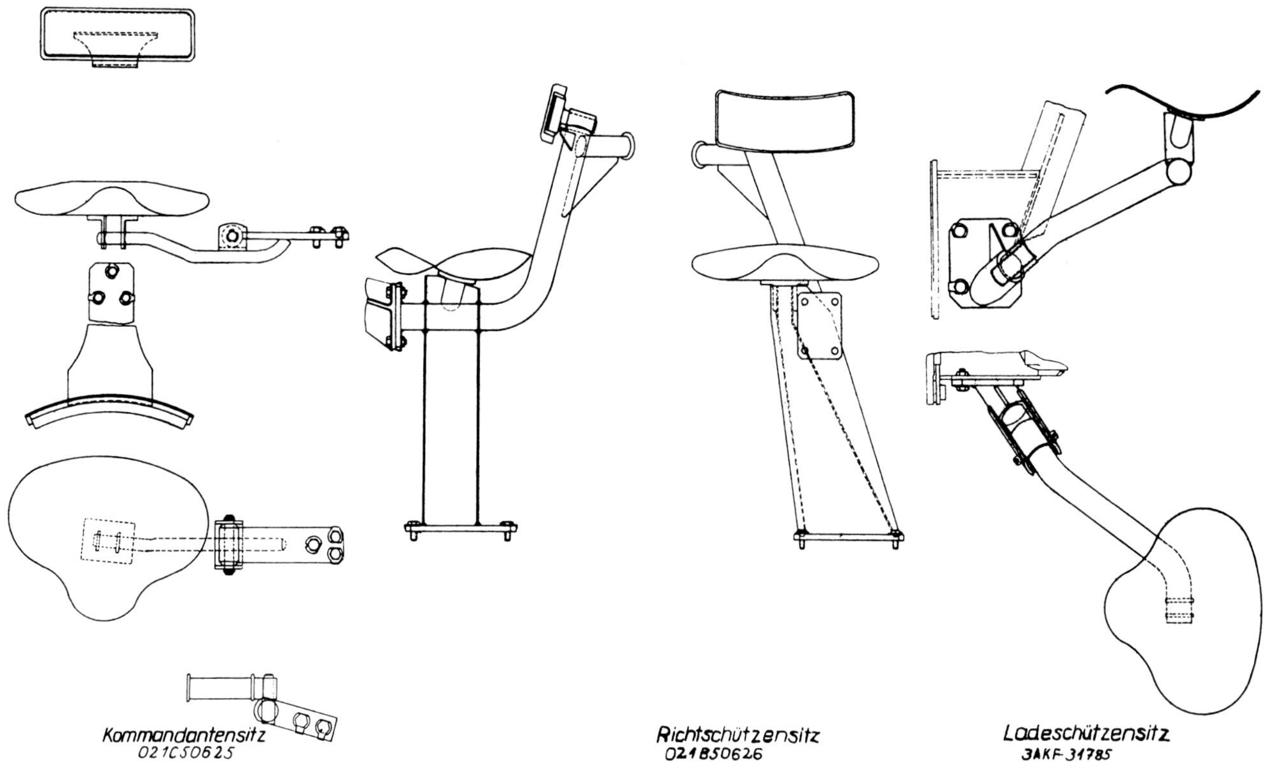

Kommandantensitz (021 C 50625) commander's seat, **Richtschuetzensitz (021 B 50626)** gunner's seat, and **Ladeschuetzensitz (3AKF31785)** loader's seat. One foot rest for the commander was bolted to the turret ring; the second was welded to the back of the gunner's seat.

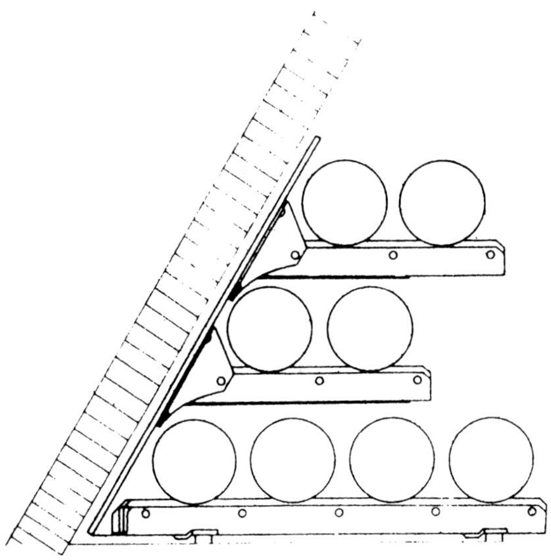

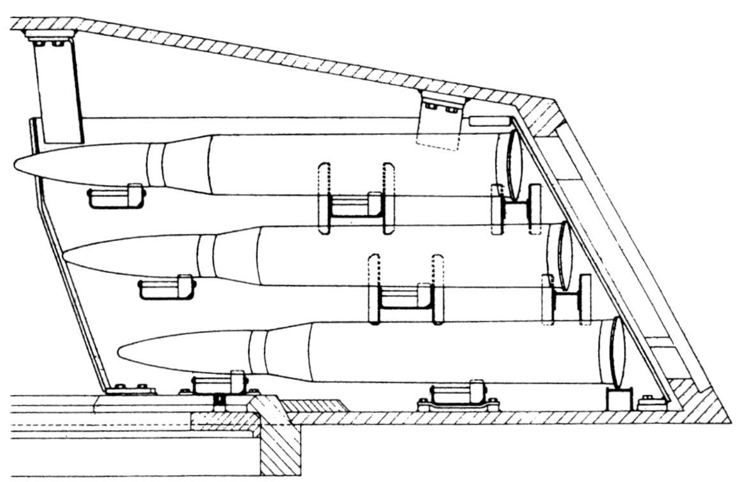

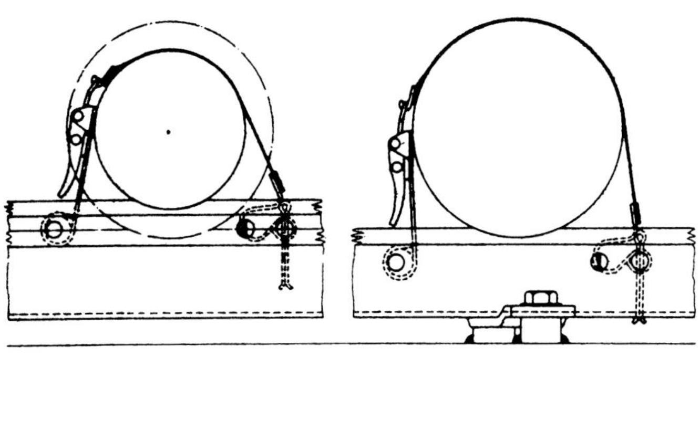

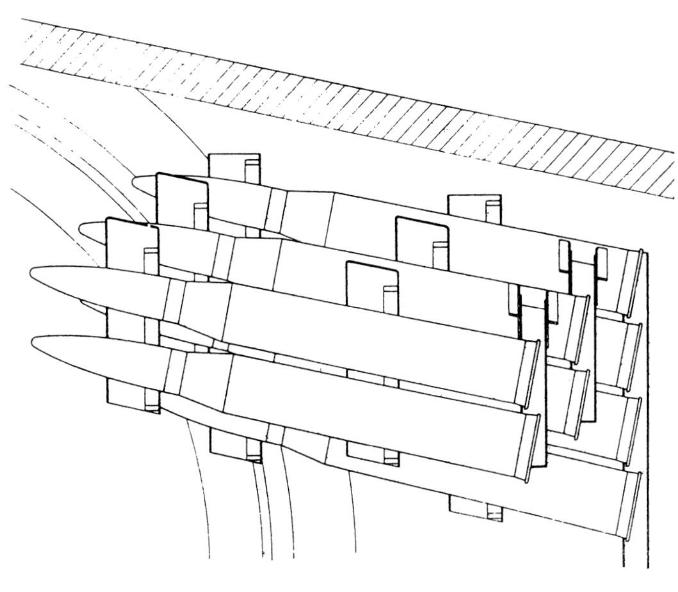

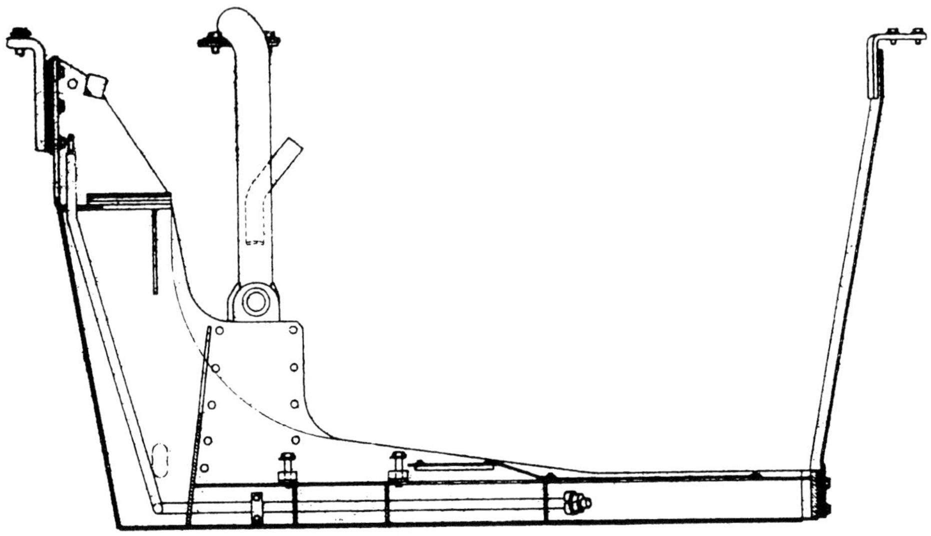

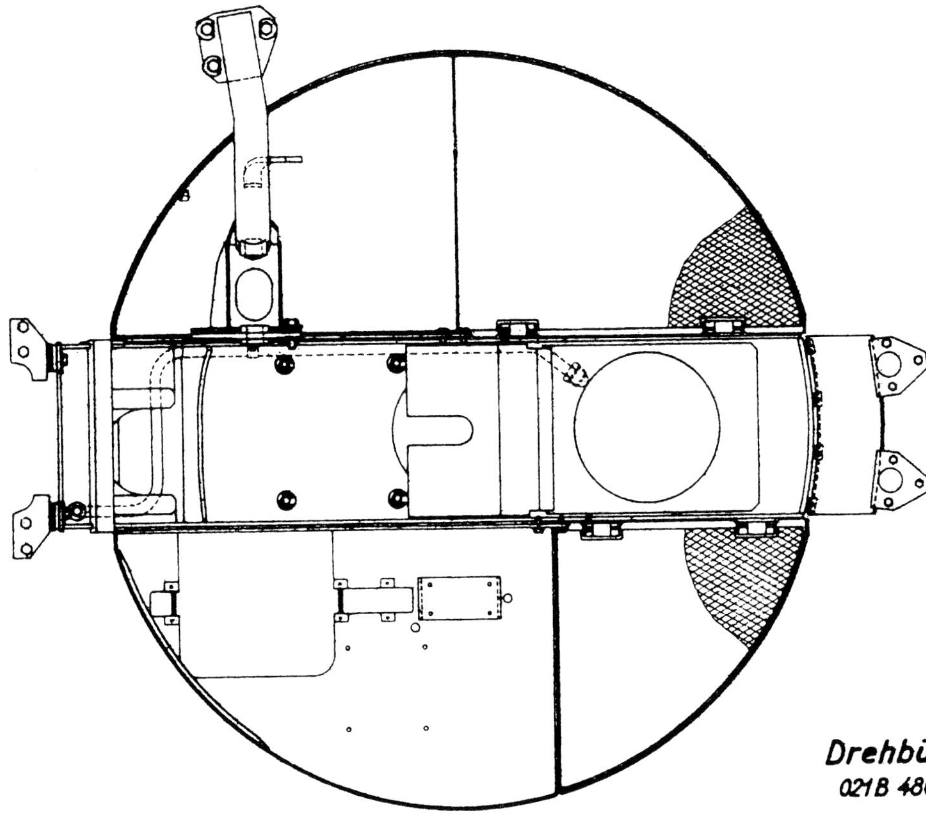

ABOVE AND ABOVE RIGHT: Side, top, and rear views of the **Drehbuehne (021 B 48021)** traversing turret platform. The platform was suspended fore and aft by traverse-drive carriers bolted to the turret ring. The gun elevation gear was mounted on the support bolted on the right side. Floor panels at the rear were hinged for access to the batteries. The square cutout in the floor was for a foot pedal to control the power traverse.

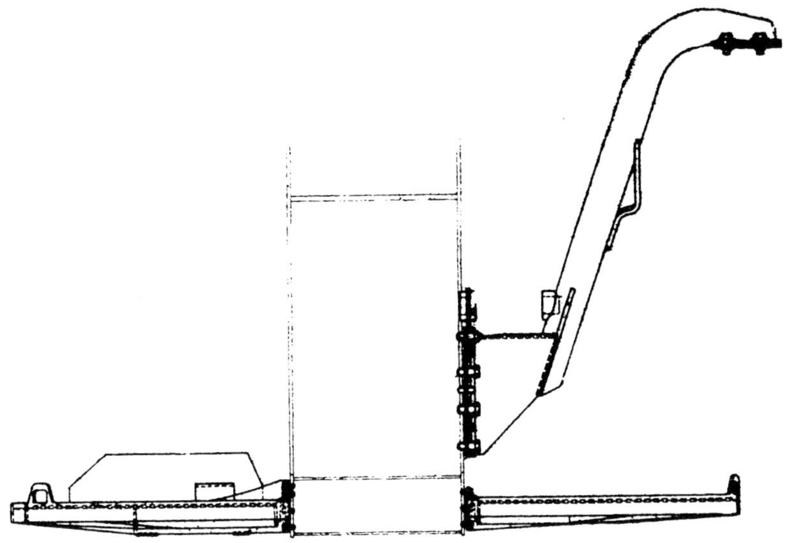

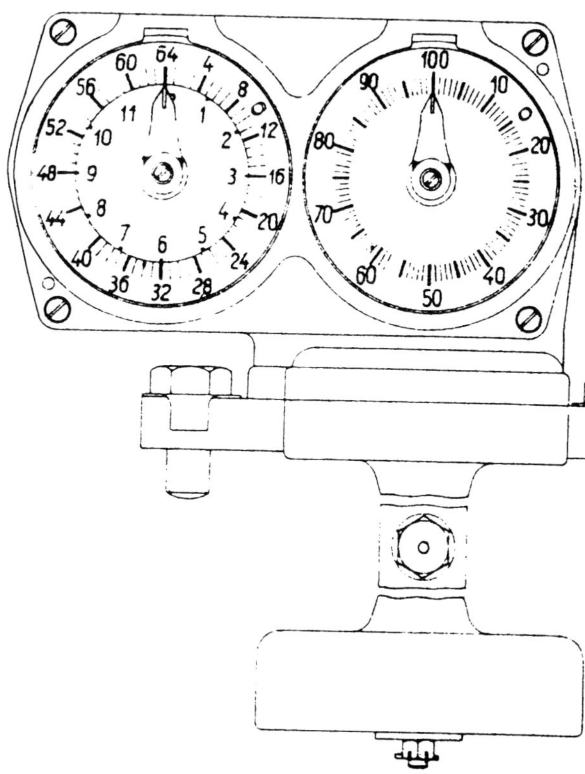

The **Zwoelfuhrzeiger (021 B 48030)** azimuth indicator mounted on and driven by the turret ring had dials for percent, mils, and hours. In combination with the **Zwoelfuhrzeigerring** (rotating azimuth indicator ring in the commander's cupola), it was to be used by the commander to direct the gunner onto selected targets.

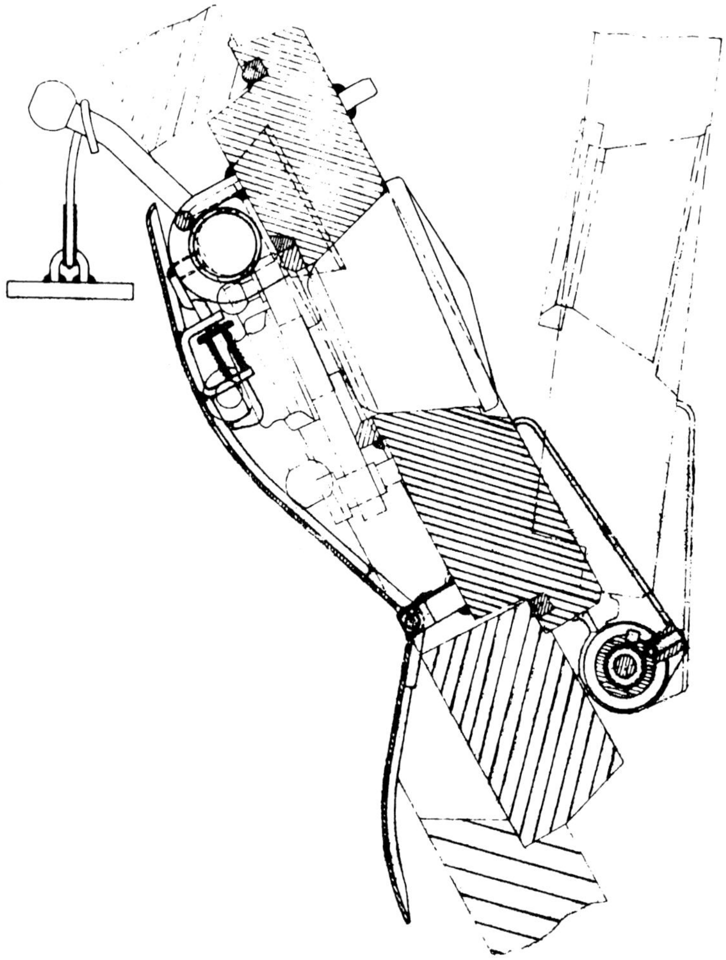

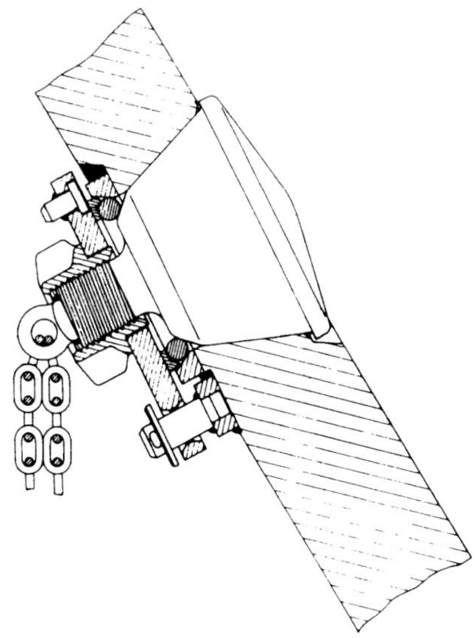

ABOVE AND BELOW: The **MP-Stopfen (021 B 48031)** pistol port plug was locked in place by a pivoting collar. Two chains were fastened to the plug: the shorter chain to hold the ejected plug, the longer chain to pull it back into place.

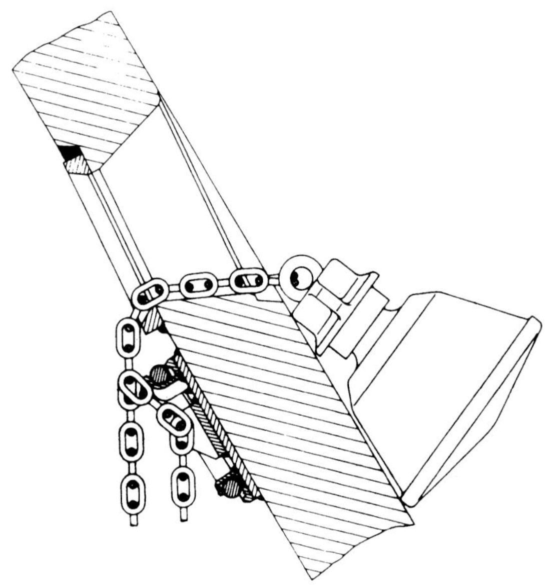

ABOVE AND BELOW: A cross-section view and inside view of the **Lukendeckel, hinterer (021 B 48017)** rear turret hatch showing details of the torsion bars, locks, sheet metal slide, and pistol port.

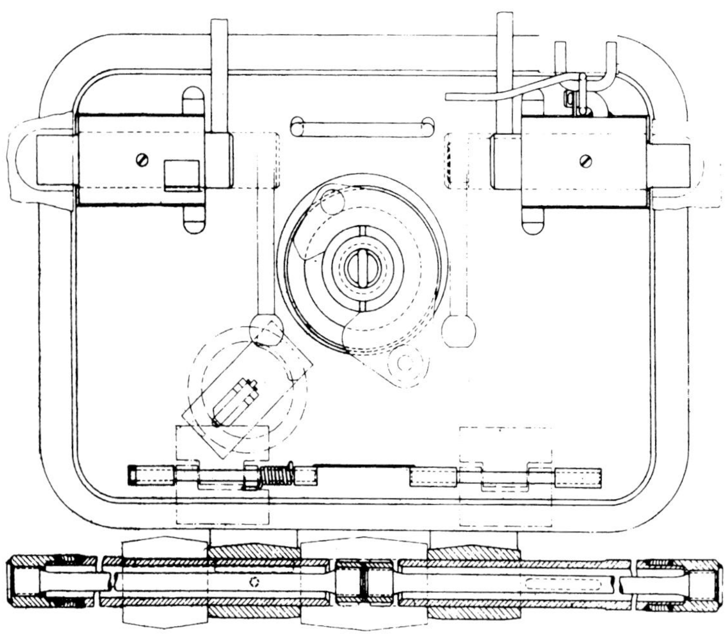

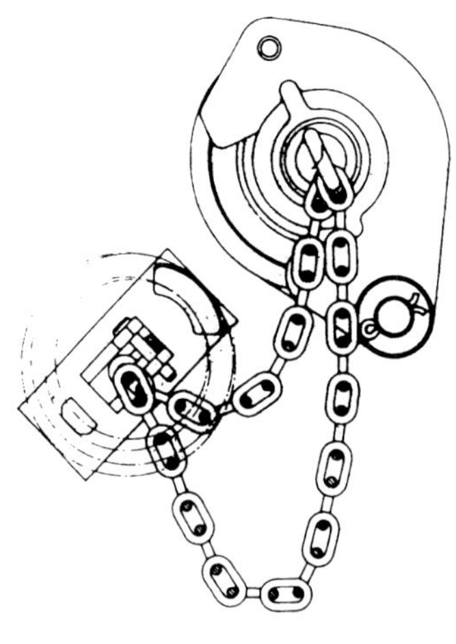

6.2 TIGER II PRODUCTION

In early October 1942, plans for production of the **VK 45.03** were reviewed. It was considered intolerable that such a large number as 424 **VK 45.01 (H)** needed to be produced before Henschel could start **VK 45.03** production. Henschel recommended that they produce 330 **VK 45.01 (H)** and then switch production to 170 **VK 45.02 (H)**, so that **VK 45.03** production would start at number 501. Oberst Thomale (with the **Panzerkommission**) rejected the plan to insert a **VK 45.02 (H)** between the **VK 45.01 (H)** and **VK 45.03**. He consented to starting **VK 45.03** series production in September 1943 and agreed to this providing 100 Tiger H3 for the Spring offensive in 1944.

In a meeting on 12 October 1942, Oberbaurat Kniepkamp (**Wa Pruef 6**) related that: *The Tiger II is not going to be produced. Start of the **Tiger III (VK 45.03)** is requested for July 1943 but is probably not possible until September 1943.*

Henschel was awarded contract number SS 006-6362/42 by **Wa Pruef 6** to assemble three **Versuchs-Fahrgestell** (trial chassis) with **Fgst.Nr. V1**, **V2** and **V3**. Their first mass production contract, number SS 4911-210-5910/42, was awarded in October 1942 by **Wa Rue (WuG 6)** for assembly of 176 **Fahrgestelle** with **Fgst.Nr.280001-280176**. Before the first chassis had been assembled for trials in October 1943, contract extensions had been awarded to raise the total order to 1234.

Following delays which Henschel blamed on having to incorporate components from the Panther II design, the first Tiger II (**Fgst.Nr.V1**) was completed by Henschel in October 1943 and accepted by the **Heereswaffenamt** inspectors in November 1943. Further delays in meeting the production goals were blamed on start-up problems. The actual production achieved for the period from October 1943 to May 1944 amounted to only 38 (compared to total production goals of 191 Tiger II for this period). Instead of having the 100 Tiger II needed to complete two **schwere Panzer-Abteilungen** ready for action in the Spring of 1944, there were only five Tiger II with combat units on 1 June 1944 (and these were mechanically unsuitable for frontline employment).

Henschel finally solved most of the start-up problems and were meeting and exceeding their monthly production goals when they were hit by a series of five bombing raids on 22, 27, and 28 September and 2 and 7 October 1944. A total of 2906 tons of high explosive and 1792 tons of incendiary bombs were dropped, the Henschel plant being the intended target. These raids destroyed 95 percent of the total floor area of the Henschel plant. Another bombing raid on the plant on 15 December further delayed recovery. In addition, heavy area bombing raids on Kassel and vicinity on 22/23 October, 30 December and 1 January resulted in further disruptions to Tiger II production.

In his report dated 31 December 1944 on the production of tanks beginning in 1940 up to 31 December 1944, Dr. Blaicher, the Chairman of the **Hauptauschuss Panzerkampfwagen** (Main Committee Armored Fighting Vehicles) under the Speer's Ministry for Armaments and War Production, left behind a realistic picture of the problems then facing the industry:

<u>*Discussion of the important difficulties in the year 1944*</u>

While during 1943 the influence of hostile air attacks was not yet very noticeable in the tank industry and the more important suppliers and contractors, in 1944 there was no single tank-producing plant which did not suffer directly and above all indirectly to a considerable extent.

Insufficient supply of supplied parts, particularly aggravated by the longish transport bans, must be counted as an additional difficulty for all firms during the whole year, but especially in the last four months.

<u>*Damage inflicted through enemy attacks on tank-body firms*</u>

*Fried.Krupp, Essen: Through the attack of 25 October 1944, the production of **Pz.Kpfw.IV** armor components in particular was hindered considerably. Production in the Tiger workshops and rolling mills was also considerably obstructed by continuous attacks on Essen.*

<u>*Damage to the AFV plants*</u>

Tiger production at Henschel, Kassel: Extreme difficulties have been caused by the somewhat complicated power supply and by the labor situation through repeated air attacks on Kassel. Production seriously obstructed by three severe attacks in September 1944 and three further attacks causing a long-term stoppage of power.

<u>*General disturbing influences*</u>

Considerable loss of man-hours through air-raid warnings was added to these difficulties in the various districts. Individual statistics are not available at the moment. MIAG, Braunsweig, for instance, lost about 300,000 man-hours through direct and indirect effects of enemy attacks and through warnings in October 1944. Considerable dispersion was undertaken by all firms to avoid total losses through these enemy attacks. A further deterioration of the development of production, and especially the catastrophic aggravation of the transport situation, became a disturbing factor to this policy of dispersing stores and dumps and the supply of components with the production plant as an organic whole.

It should be especially stressed that the internal factory conditions created by the effects of the enemy attacks brought with them an extraordinary strain on the managerial and administrative staff, and that the extremely large reduction of German labor, which is decisive in dealing with the catastrophes, has badly retarded reconstruction and the overcoming of the difficulties during the current year. When repeated and regular production breakdowns occurred, which in some plants lasted the whole year round, the changeover from a state of improvisation to an organic condition of production could hardly be achieved, in spite of the most strenuous efforts of the management. The extraordinary extent of the stoppages throughout the whole of the armaments industry caused all external assistance in overcoming these difficulties to diminish more and more, and the firms were largely dependent on helping themselves.

Added to these local difficulties was the large number of stoppages of deliveries on the part of suppliers and subcontractors. The production situation was made extremely difficult by the severe interruptions of communications and the disruption of the supply of components which became increasingly worse. This was caused by the transport situation, which is deteriorating on a catastrophic scale.

At this moment one can no longer speak of planned production at all.

On principle, I would like to mention that tank production has to deal with a maximum of difficulties such as no other branch of production can boast, by reason of the target demanded on the one hand, and on the other hand through the addition of all the difficulties of the branches of other main committees connected with it. To mention that a Panther possesses 26,000 parts will be sufficient to prove this statement.

In spite of all these difficulties, the factories have grown with their tasks, and the plant managements, as well as the whole organization, are trying to adapt themselves as best as possible to these difficulties in order to overcome them and to obtain the best possible results for German armaments.

But the pessimism of Dr. Blaicher wasn't shared by **Wa Rue (WuG)** in their projections dated 30 January 1945 in which they predicted that production could recover up to 125 Tiger II per month through December 1945, as follows:

	Jan	Feb	Mar	Apr	May	Jun	Jul	Aug	Dec
Tiger II	40	35	45	55	90	105	115	125	125

However, reality soon returned with the release of the **Panzer Notprogramm** (Emergency Panzer Production Program) dated 1 February 1945, as follows:

Type	Firm	Feb	Mar	Apr	May	Jun	Total
Tiger II	Henschel	50	60	70	70	47	297
Tiger II	Ni-Werk	-	-	13	40	-	53

This emergency program called for production to cease after completion of an additional 350 Tiger II (for a total of 770). This is the also the earliest document found that reveals plans to involve Nibelungenwerk in the assembly of Tiger II chassis.

On 21 February 1945, Dr. von Heydekampf approved the following program for ending Tiger production:

	Feb	Mar	Apr	May	Jun	Jul	Aug	Sep	Oct
Henschel	45	50	50	60	60	60	60	45	-
Ni-Werk	13*	40*	37*	25	25	25	25	-	-

*Jagdtiger

In this document Henschel reported: *Up to the end of January 1945, 417* ***Tiger II Serienfahrzeuge*** *and 3* ***Versuchsfahrzeuge*** *as well as 60* ***Jagdtiger*** *have been produced. The total number in the final production program is increased to 950 Tiger II and 150* ***Jagdtiger*** *including a further 530 Tiger II and 90* ***Jagdtiger****. Henschel is giving Nibelungenwerk a contract to assemble 100 Tiger II chassis.* [The last report on production from Nibelungenwerk dated 29 April 1945 stated their production achievements in April and reveals that they were still attempting to complete **Jagdtiger** and **Pz.Kpfw.IV** in May 1945. But, there is no mention of a single Tiger II being completed at Nibelungenwerk.]

As a result of Allied troops overrunning Kassel and vicinity, Henschel had ceased all tank production by the end of March 1945. The Allies' bombing campaign had caused the loss in production of at least 657 Tiger II (940 originally planned versus 283 actually produced) during the period from September 1944 through March 1945.

6.2.1 Henschel & Sohn, Kassel

Henschel & Sohn in Kassel were the only firm that assembled Tiger II chassis in their Werk III in Kassel-Mittelfeld. 8000 workers were engaged in tank production, working in two shifts each of 12 hours. The night shift was stated to have only 50% of the output of the day shift. A pure **"Takt"** (assembly station) system was employed in the assembly shop, consisting of 9 **"Takte"** (assembly stations) each of 6 hours. The total time estimated to complete a tank, including the various machining processes, was said to be 14 days. An average of 18 to 22 tanks were carried in the hull assembly shop and 10 on the final assembly line.

The part of the Mittelfeld plant on the left-hand side of the main railway line, looking north, was concerned with truck rework including engines, and the manufacture of locomotive components. The main tank component stores were located here, as were sheds containing tank turrets and hulls. The second part of the Mittelfeld works (i.e., to the right of the main railway looking north) consisted of four shops numbered 1, 2, 3, and 5. Number 4 Shop, which was supposed to have been constructed some time ago, was in fact never built. Shop Number 1 was a general machine shop mainly concerned with locomotive components. Shop Number 2 was the same. Shop Number 3 was concerned with machining of various tank components such as suspension arms, steering mechanism, hatch cover plates, etc. and hull assembly. Shop Number 5 was divided into two parts. On the left (looking north), it was engaged in locomotive work and on the right, with final assembly of tanks, i.e. fitting of the turrets (received in a finished condition from Wegmann) to the already completed hulls.

Two hull assembly lines were fitted in Shop Number 3, one on each side of the two large bays. The remainder of each bay was devoted to the machining of suspension arms, hatch cover plates, steering mechanism, top decks, and assembly of the resilient bogey wheels. The hull arrived as a welded unit with holes for suspension arms, final drive, rear idler, etc. rough bored. It then went to the 8-spindle horizontal borer (in fact only 6 were being used), where the holes in the hull sides for the suspension arms were finish-bored. The next process is the 4-spindle borer which finished the holes, both front and rear, for the final drives and idlers. The hull then went to a vertical lathe for machining the face on the superstructure roof for the turret ring. Simultaneously with this process, the milling of the hull sides to receive the final drive casing was carried out. The hull then moved on to an assembly line so that the various components could be added.

Significant differences between the two hull machining lines were due to the fact that, whereas in the one case the hull was mounted straight away onto bogeys running on rails (so that accurate alignment was thereby maintained with the help of adjustable jacks for each operation), in the other case (the older line), the hulls had to be craned from one drill jig to the next, necessitating accurate alignment before each operation. In this case, the hull moved sideways, whereas in the former case they were positioned nose to tail. The machining of the superstructure top plate to receive the turret ring, and the end milling of the hull side plates to receive the final drive casing, were registered for all subsequent machining operations. The process therefore had to be carried out first on the newer machining line if satisfactory results were to be obtained, because the hull was adjusted for alignment only once. On both assembly lines, the hulls moved nose to tail and were carried on bogies mounted on rails.

6.2.2 Wegmann Waggonfabrik A.G., Kassel

This was the only firm engaged in the final assembly of turrets for the **Pz.Kpfw.Tiger Ausf.B**. The plant employed some 1200 workers. Its optimum number was given as 1500. They worked in two shifts of 10 hours which was increased to 11 hours toward the end of certain months in an endeavor to reach the target output figure set by the Speer ministry. The only machining processes carried out by this firm were the finishing of the welded turrets, processing and assembly of the turret ring, and fabrication of the traverse mechanism. The other components arrived at the firm in a finished state ready for assembly. The welded turret, cover plates and mantlet were furnished by Krupp and DHHV; the armament by **Heereszeugamt Nuernberg**; the hydraulic traverse motor by Boehringen, Goeppingen; the compressor by APAG, Goessnitz; the telescopic sight by Leitz, Wetzlar; the turret ring blank by Mitteldeutsche Stahlwerke A.G. and Krupp; and the turret race by Sonderring Walzlager.

With trained personnel, the complete assembly, including the machining processes, of the Tiger II turret was estimated to take 6 to 7 days, but using diluted labor (which was predominantly foreign), the average time was 10 to 12 days.

There was no rigid **"Takt"** system employed in this plant, although higher authorities had desired it. The managing director, however, maintained that such work did not lend itself to strict production control.

Processes and Plant Layout - The turret assembly was carried out in the following stages:

a. The turret was placed upon a bogey stand and external brackets for spare track links, lifting hooks, etc. were welded on.
b. The ring of the commander's cupola was welded onto the turret roof.
c. The turret was inverted and the stowage accessories fitted.
d. The turret was set up on one of the four special vertical lathes for the machining of the turret face and inner diameter to receive the turret ring and ball race.
e. The turret was removed from the lathe and the ring and ball race fitted. The turret ring was machined concurrently in a separate shop.
f. The turret was then placed upon the bogey stand in its normal position and the ammunition container was installed in the rear bulge.
g. A separate bay was devoted to receiving the main armament completely assembled with gun cradle, recoil mechanism, and muzzle brake. This was stripped down and cleaned and the gun and coaxial MG checked for alignment.
h. The gun cradle was installed in the turret together with the equilibrator. A graduated wooden beam and weight were employed in order to set the equilibrator.
i. The main armament, together with the coaxial MG and gun sight, was installed.
j. Next the turret platform was fitted.
k. The traverse gearbox and hydraulic power unit was installed.
l. The elevating mechanism was fitted.
m. Electrical wiring and final stowage fitting completed the assembly.

6.2.3 Component Production

6.2.3.1 Armor

Fried.Krupp A.G. in Essen was the primary fabricator of armor components for the Tiger II. They were awarded contract number SS 210-5813/42(H) dated 14 November 1942 for 259 armor hulls and turret bodies for the **Pz.Kpfw.Tiger(Henschel) Ausf.3.Serie (= Tiger H3)**. Contract extensions had been awarded by March 1943 to bring the total up to 1036 armor hulls and turret bodies, and by November 1943 to 1493 (in serial number series 280001 to 281493). The last contract increase to SS 210-5813/42 was awarded to Krupp on 19 August 1944 for an additional 107 armor hulls (serial numbers 284001 to 284107) and 99 armor turret bodies (serial numbers 284001 to 284099).

Including the 50 armor turret bodies produced for the **VK 45.02(P)** which were converted for mounting on the **VK 45.03**, Fried.Krupp A.G., Essen manufactured 444 armor hulls and 385 armor turret bodies by the end of February 1945.

Two additional steel firms, Dortmund-Hoerder-Huetten-Verein (DHHV) and Skoda, also fabricated armor hulls and armor turret bodies for the Tiger II. They started later than Krupp, with DHHV completing only six hulls and six turrets and Skoda only four hulls during the period from February to May 1944. At the end of the war, DHHV claimed to have produced 157 armored hulls and turrets for the Tiger II plus 35 turrets for Tiger II hulls completed by Skoda.

6.2.3.2 Engines

Maybach, Friedrichshafen, the inventor of the high performance **Maybach HL 230 P30 Motor** didn't maintain statistics on the production of the **HL 230** separate from its predecessor the **HL 210**. Altogether in their **HL 210/230** series, Maybach produced 153 in 1942, 4346 in 1943, 2973 in 1944, and 987 in 1945. After production of an estimated 1785 HL 230 motors from January through April 1944, Maybach was completely knocked out of engine assembly by a bombing raid in late April 1944. Destruction was so devastating that Maybach didn't get back to producing the **HL 230** until October 1944.

Auto Union, Chemnitz at their Siegmar-Werk produced and delivered 4366 **HL 230** engines for the Panther and Tiger in 1944 and 1945. Having produced 800 engines each month in July and August, they were hit by bombing raid on 11 September 1944, dropping production to 198 in September and 312 in October. Auto Union delivered 332 **HL 230 P30** motors to Henschel for installation in the Tiger II in 1944 and 1945.

6.2.3.3 Transmissions

Adlerwerke, Frankfurt am Main was the main assembly firm for the Maybach-designed **Olvar 40 12 16 B Getriebe** installed in the Tiger II. Statistics were not maintained separately for the production of the A and B variants of the **OG 40 12 16** transmissions but were reported together as 130 in 1942, 875 in 1943, 819 in 1944, and 26 ending in March 1945, sequentially numbered from 1 through 1850. A second factory run by Zahnradfabrik Friedrichshafen A.G., Waldwerke Passau, started production of the **OG 40 12 16** transmissions in early 1944 and had completed serial number 20117 by September 1944.

6.2.3.4 Steering Gear

Henschel & Sohn, Kassel was the inventor and sole producer of the double-radius **Lenkgetriebe L801** steering gear for the Tiger II. Production was in a single series, sequentially numbered starting with 1.

6.2.3.5 Main Gun

All of the gun tubes for the **8.8 cm Kw.K.43 L/71** were fabricated at a single steel works, DHHV. However, final assembly of the operational guns, complete with cradle, gun shield, and telescopic sight, was dispersed, with 55 percent completed in Frankfurt and 45 percent in Dortmund. A total of 802 **8.8 cm Kw.K.43 L/71** were completed, test fired, and accepted by **Heereswaffenamt** inspectors between October 1943 and the end of February 1945.

6.2.3.6 Telescopic Sights

Out of a total order for 360 **T.Z.F.9b/1** binocular telescopic gun sights, Ernst Leitz G.m.b.H., Wetzlar reported that they had delivered 62 in January 1943. A total of 3000 **T.Z.F.9d** monocular telescopic gun sights was ordered from Leitz, of which 1092 were produced by the end of March 1945 and 989 were delivered by the end of February 1945.

TIGER II PRODUCTION

	Highest Monthly Goal	Production Goals Reset Each Month	Actual Production Accepted by WaA	Approx. Fgst.Nr. at End of the Month
Sep43	1			
Oct43	3			
Nov43	5	1	1	V1
Dec43	8	2	0	V3
Jan44	12	3	5	280003
Feb44	25	5	5	280008
Mar44	37	6	6	280014
Apr44	50	12	6	280020
May44	50	20	15	280035
Jun44	50	25	32	280067
Jul44	65	45	45	280112
Aug44	100	80	94	280206
Sep44	100	100	63	280269
Oct44	120	120	26	280295
Nov44	140	40	26	280321
Dec44	140	60	56	280377
Jan45	140	60	40	280417
Feb45	150	35	42	280459
Mar45	150	45	30	280489

TIGER II COMPONENT PRODUCTION

	Armor from Krupp		Turrets from Wegmann	8.8cm KwK43 L/71	TZF9d from Leitz
	Hulls	Turrets			
Oct43			0	2	
Nov43	8*	5*	1	11	
Dec43	4	10	2	20	
Jan44	8	6	3	12	
Feb44	12	11	5	1	
Mar44	21	15	6	27	0
Apr44	27	15	11	39	10
May44	43	27	15	38	26
Jun44	↑	↑	33	71	35
Jul44	176	159	44 est.	80	77
Aug44	↓	↓	90	102	130
Sep44	72	82	73	113	142
Oct44	52	50	35 est.	63	185
Nov44	6	0	41	72	108
Dec44	5	0	56	67	90
Jan45	4	0	32	22	118
Feb45	6	5	46	62	120
Mar45	??	??	??	??	51

* Number produced by the end of November 1943

Stand des Programmes „Pz" im Fahrzeugbau 1944

Benennung		Jan.	Febr.	März	April	Mai	Juni	Juli	Aug.	Sept.	Okt.	Nov	Dez.
Fahrgestell Tiger I mit Aufbauten	soll	93	95	95	95	95	75	21 10 45 61	9 Rep				
		819	914	1009	1104	1199	1274	1340	1349				
	ist	93	95	86	104	100	75	64	6				
		819	914	1000	1104	1204	1279	1343	1349				
Fahrgestell Tiger II mit Aufbauten	soll	3	5	6	12	25	30	45	100	120	120	140	140
		6	11	17	29	54	84	129	229	349	469	609	749
	ist	3	5	6	6	15	32	45	97				
		6	11	17	23	38	70	115	212				

An original report from Henschel for 1944 showing the **Soll** (planned) and **Ist** (actual) production along with the cumulative total for each month as the Tiger I was phased out and the Tiger II was phased in.

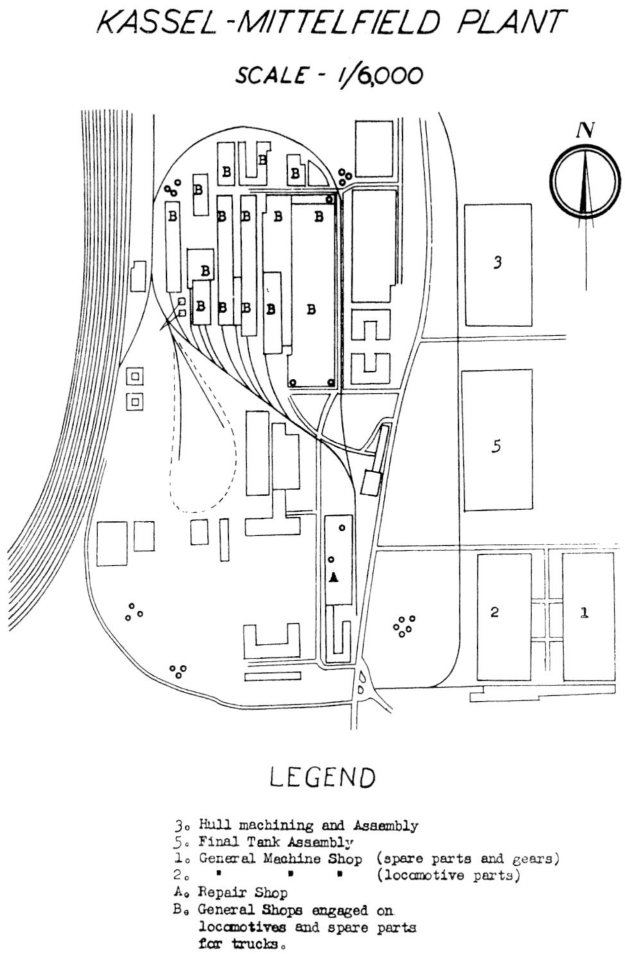

Layout of the Henschel Werk III assembly plant.

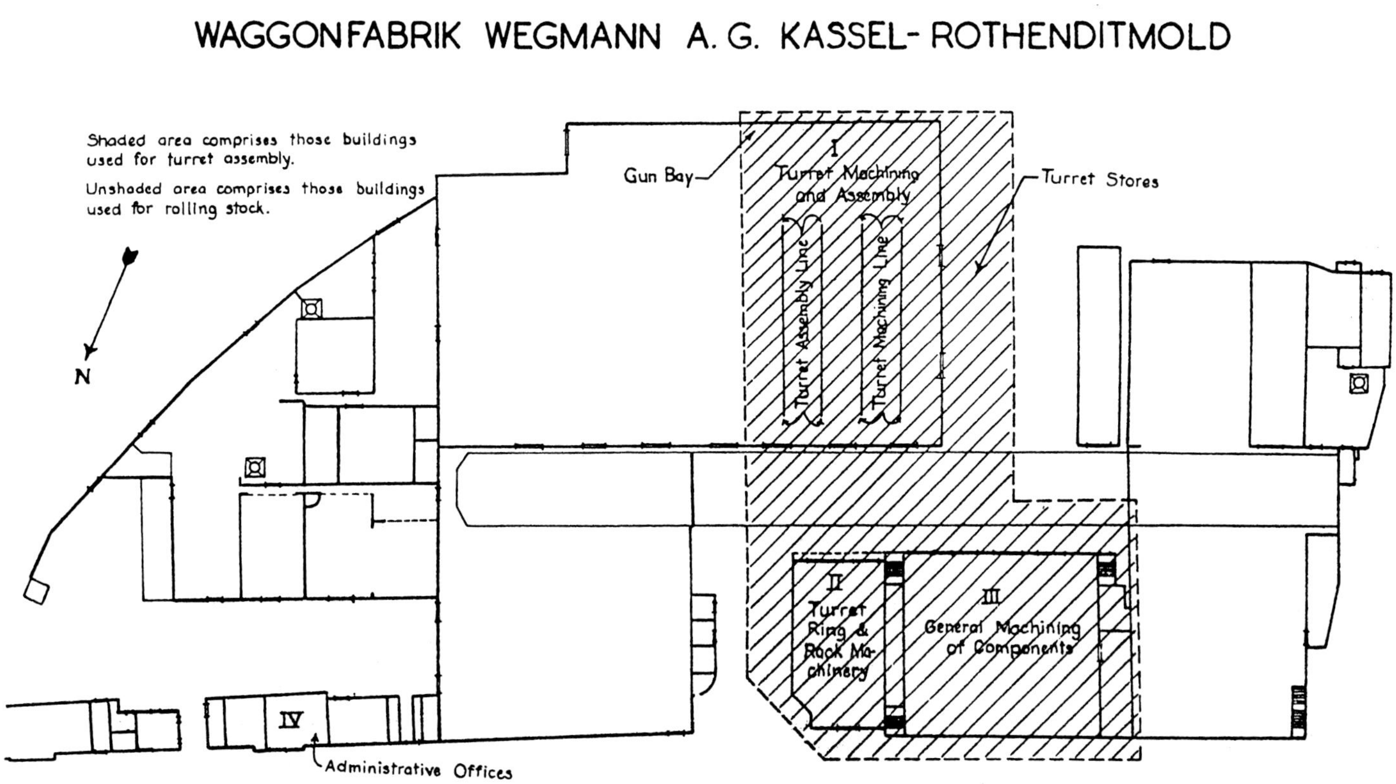

Layout of the Wegmann assembly plant.

6.3 MODIFICATIONS DURING PRODUCTION

As with all production series German Panzers, modifications were frequently introduced during the production runs. These modifications were prompted by a desire for:

- improved automotive performance,
- increased firepower,
- added protection,
- simplified design for easier manufacturing,

or were forced by shortages.

The modifications are grouped as automotive, turret and weapons, or camouflage. They are listed in the chronological order in which the changes first appeared on completed Tiger II leaving the assembly plants. In rare cases, several months elapsed between the first appearance of a modification and the time that it was present on all newly produced Tiger II. This was due in part to "first in, last out" tendencies when older stockpiled parts were covered, buried, or made inaccessible by storing shipments of newer parts. The newer parts, being easier to obtain, were used first until their removal allowed access to the older parts.

6.3.1 AUTOMOTIVE PROBLEMS

In his **Anmerkungen zu Tiger B** written in February 1945, Dr.Ing. Aders, head of Henschel's design office, recorded the source of and solution for many of the automotive problems:

In February 1943, ***Wa Pruef 6/III*** *required thorough compatibility between the Tiger B and the Panther II that was also in development by M.A.N., Nuernberg. The highest possible number of complete components were to be exchangeable. Meetings were held and a designer from Henschel worked for a while at M.A.N. What was achieved was that the following components designed for the Tiger B were taken over by the Panther II: L801 steering unit, the final drives, and the Argus brakes. The Tiger B had to take over from the Panther II the engine cooling system, engine compartment, transmission ventilation, fuel system, ventilation for the engine exhaust pipes, engine compartment deck, engine exhaust system, and turret hydraulic drive. These changes caused the completion of the design to be delayed by about three months.*

*The inherited components taken over in this way turned out to be disastrous, in spite of insurance by Dipl.Ing. Jaeger (****Wa Pruef 6****) specifying that only tested and proven designs be used. Of course, deficiencies in the cooling system could be corrected, but the rest left behind mistrust by Henschel & Sohn. The torque-limiting clutches from Argus had design faults allowing a lot of slippage and therefore diminished the effectiveness of the cooling system.*

The opinion of Henschel's assembly shop is that the ***HL 230*** *engine layout has very much better attributes and is better developed for assembly work than the engine layout in the Tiger E.*

The numerous joints in the lines of the fuel system and the positioning of the filler tank presented problems, primarily overheating, fuel evaporation, and fires. The cause lay partially in the unsuitable seals (after going through a lot of trouble these were replaced by soft aluminum seals), partially in the standard connectors (made from ***Elektron*** *metal and damaged as soon as they were connected with steel) which were adopted from the aircraft industry, and finally in the large number of connections (about 180). The number of connections was reduced to about half by diverting from the use of standard parts and other measures.*

The fan for ventilating the transmission and cooling the exhaust pipes, proven successful in the Tiger E, had to be abandoned in favor of the layout on the rear end of the engine. Ventilation and cooling air for the transmission were then drawn by a radiator fan which turned out to be insufficient. At the present it is still not certain if increasing the cross-sectional area of the duct has solved the problem. Henschel only learned by chance in early-April 1944 that M.A.N. had solved this problem in the Panther I by air cooling with a oil-water heat exchanger in the engine cooling water reserve tank.

The categorical demand of ***Wa Pruef 6/III*** *in mid-April 1944 to install a* ***Draeger-Gasschutz-Anlage*** *(poison gas protection system) made oil cooling necessary. The* ***Draeger-Gasschutz-Anlage*** *was designed to maintain the fighting compartment with a slight overpressure, and this negated maintaining the air cooling of the transmission. Installing a 35 liter water tank to cool the transmission oil had succeeded in sufficient cooling without significant water evaporation as long as the vehicle wasn't overdriven. But because the* ***Draeger-Gasschutz-Anlage*** *had to be mounted in the location previously occupied by the transmission cooler, it couldn't be prevented that heat from the transmission was transferred to the engine coolant. An expedient solution by locating a 10 liter tank on the bottom of the hull could possibly be sufficient to transfer the heat to the armor body, but this solution would bring with it known mechanical unreliabilities that are unwelcome.*

Already with the first chassis it was evident that the rear deck over the engine compartment needed to be improved. It was redesigned to improve access to the engine and especially the magnetos.

Entirely separate from those associated with cooperation with M.A.N., problems have developed in a series of new components that were developed by other design firms.

Only a minor modification was needed in the earlier model A for the ***Olvar 40 12 16 B*** *transmission to work satisfactorily. It required only the removal of an auxiliary drive gear from inside the transmission housing. New transmission problems have occurred, probably caused by overheating, also from parts breaking.*

*There are only 2 mm of play between the track guide teeth and the roadwheels, while 4 mm of play are available on the Tiger E. This was specified by Dipl.Ing. Koehler (****Wa Pruef 6****) against the recommendations from Henschel & Sohn. Inexactness in fabricating steel-tired rubber-cushioned roadwheels and fastening them on the axles is too great to work with such little play. Additionally* ***"Staffelung"*** *(overlapping) instead of* ***"Schachtelung"*** *(boxing-in) the roadwheels was introduced. The roadwheel suspension arms are not evenly loaded, resulting in their being bent out of shape and setting at an angle to the direction of travel, different depending on whether it supported an inner or outer pair of roadwheels. Apparently to counter the attacks against the* ***Schachtel-Laufwerk*** *(boxed-in suspension), the* ***Staffel-Laufwerk*** *was proposed and its introduction specified by the* ***Heereswaffenamt****. Already, just as Henschel & Sohn expected, this appears to result in an imbalanced load on the tracks toward the inside. The track pins are bent and won't turn. This increases the track's resistance to bending and the track pins can't be pulled out. Track breakages are probably expected.*

EXTERNALLY VISIBLE MODIFICATIONS INTRODUCED DURING THE PRODUCTION RUN OF THE TIGER AUSF.B

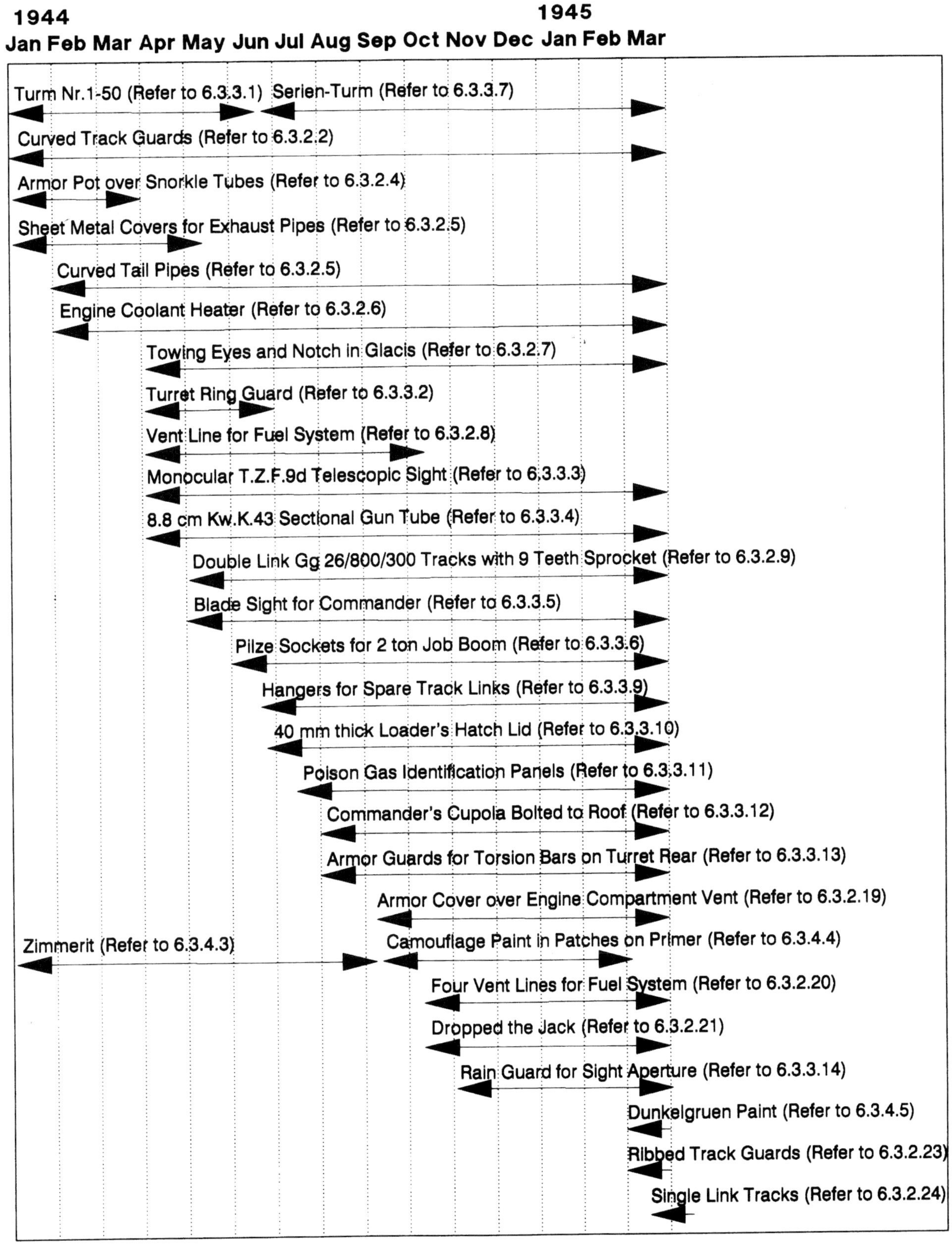

Panzerkampfwagen Tiger Ausf.B - Fgst.Nr.280001 completed in January 1944 – curved front track guards – tool stowage and **Zimmerit** – sheet metal covers for baffled exhausts – 40 mm thick plates fore and aft on the turret roof – drain beside loader's hatch.

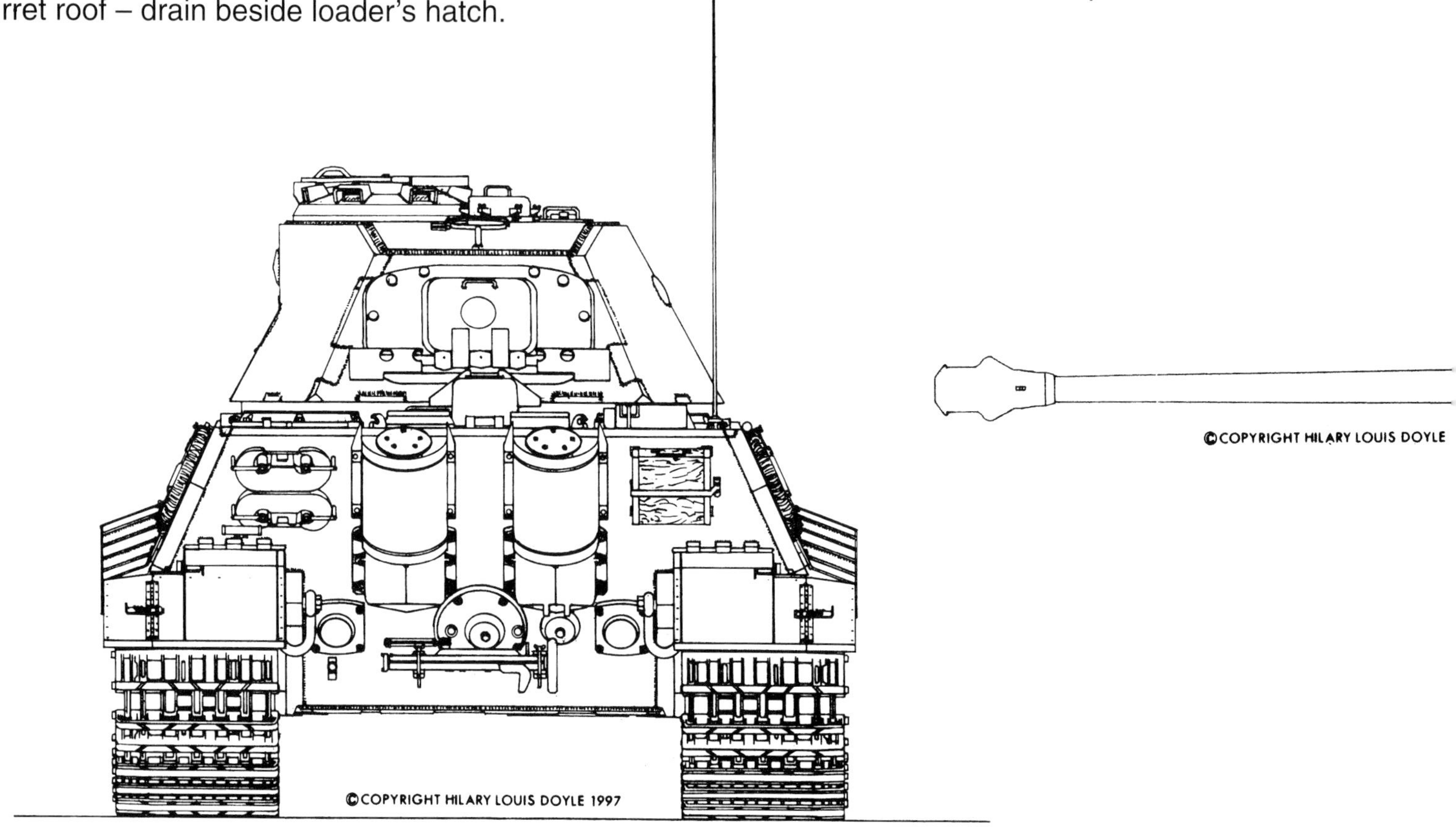

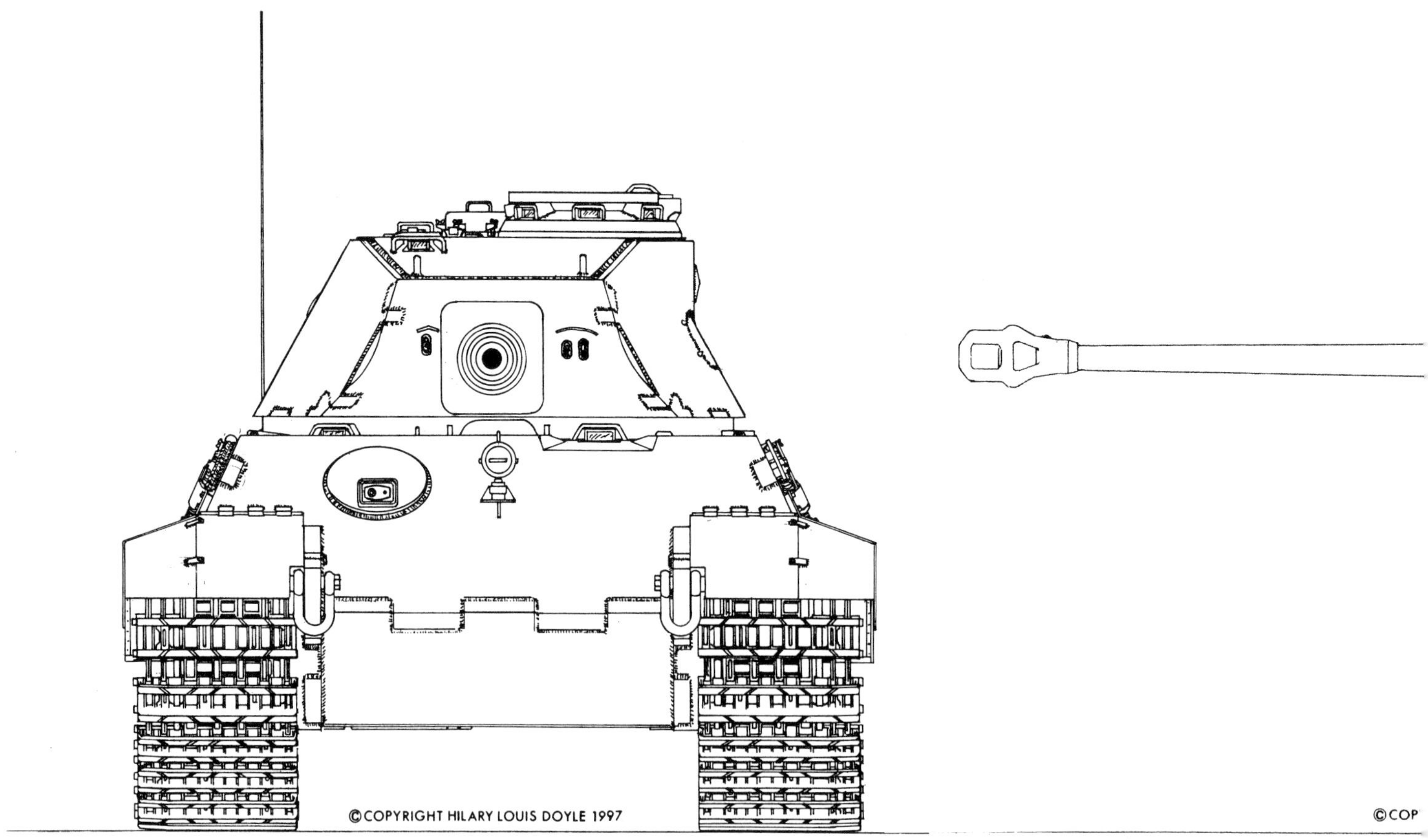

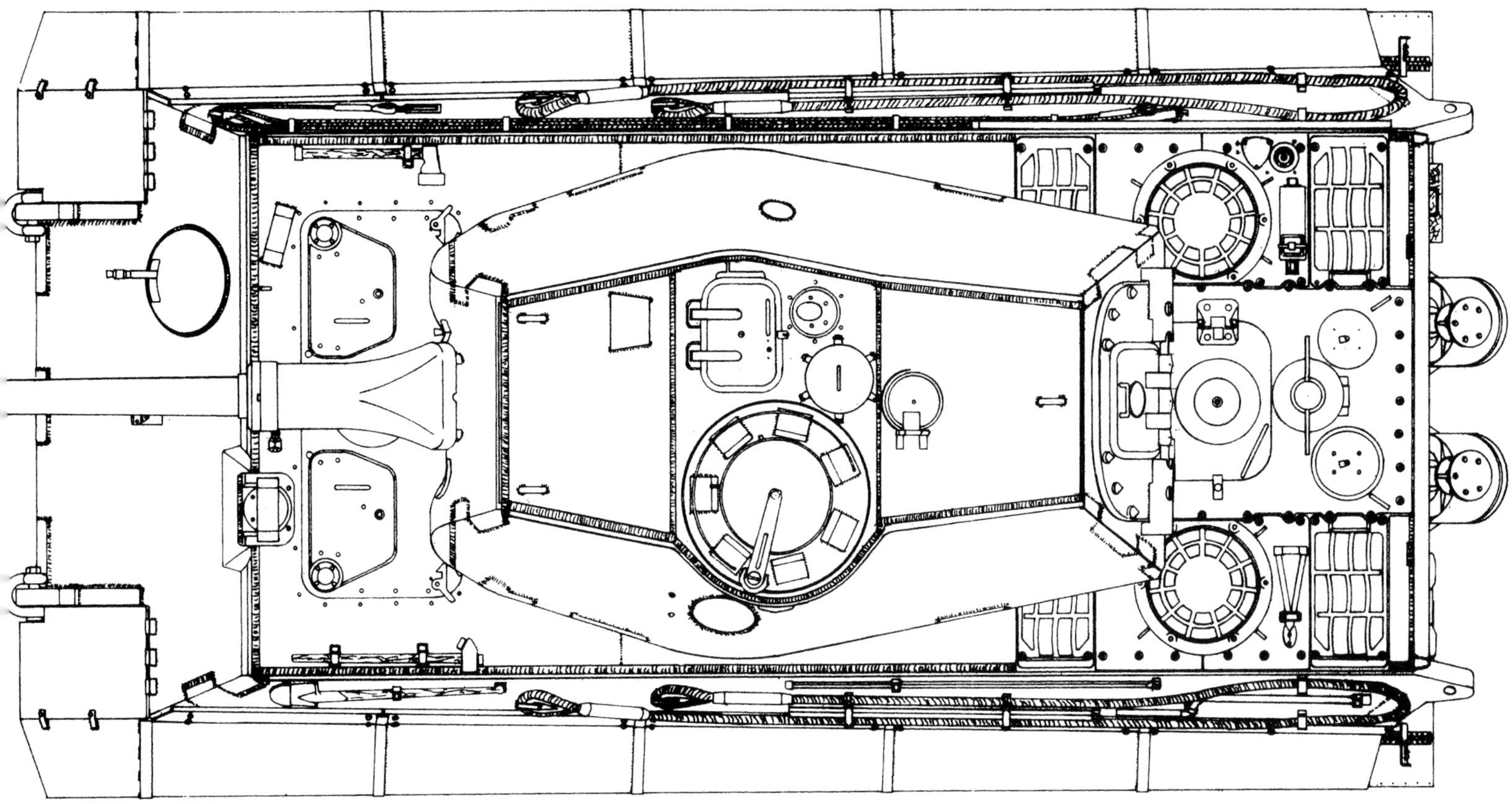

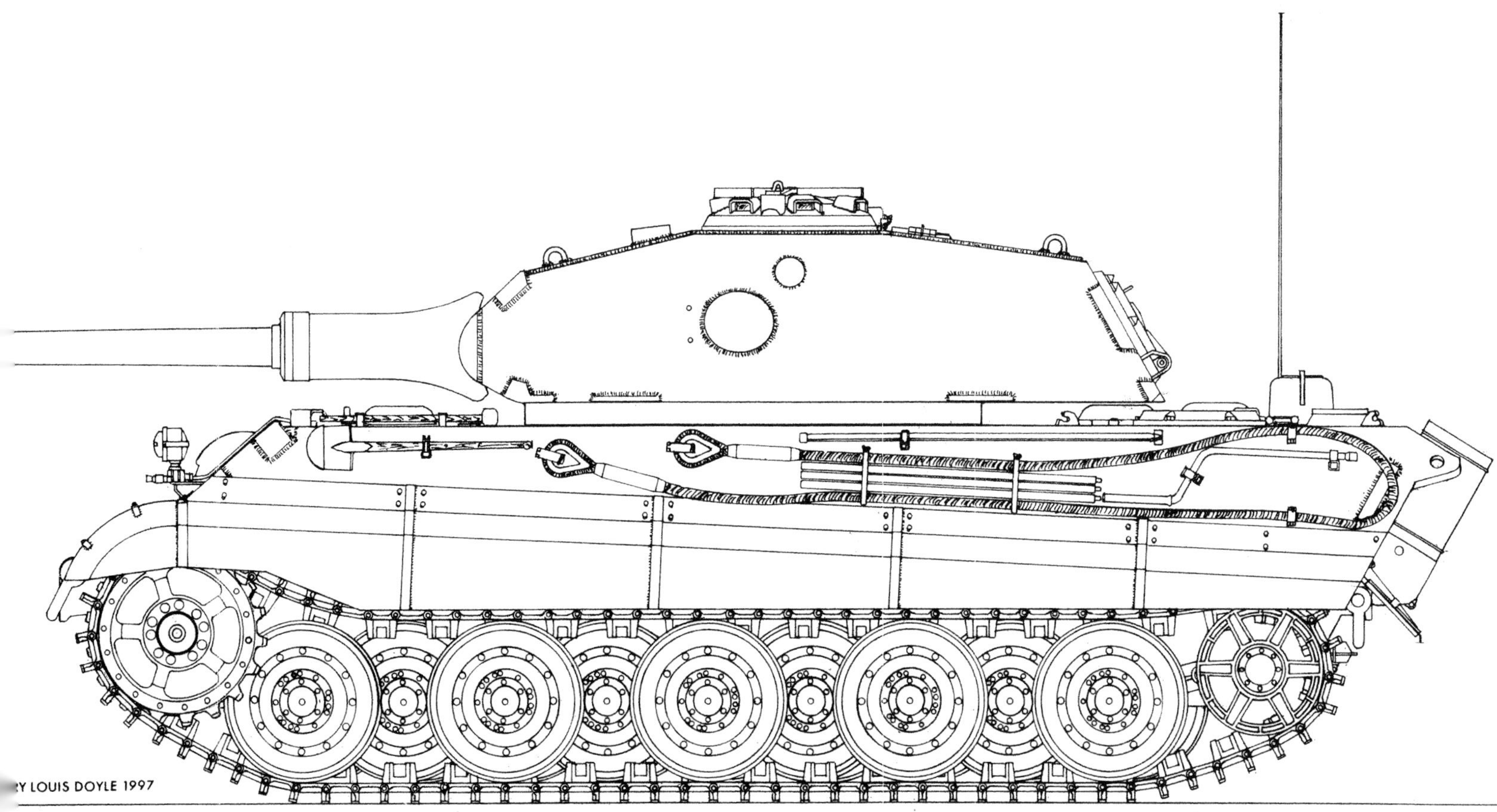
RY LOUIS DOYLE 1997

The wish for the capability to manufacture at least part of the track links by forging (in addition to achieving a standardized track for both the right and left side without differences in running resistance) led to a track made up of cast links with guide teeth and connecting links consisting of several pieces. This caused the track to be very loose, i.e., it had little stiffness to resist bending. It hasn't been determined whether this is connected with the tendency of the track to climb onto the sprocket or with imprecision of the parts. In any case, parts imprecision led to rapid wear of half of the sprocket teeth, while remarkably every second tooth remained unworn.

A new track was delivered in July 1944 in which the connecting links (without guide teeth) were cast as a single piece. This has significantly increased the firmness of the track, namely resistance against sideward forces. The firm of Miag in Braunschweig is working on a design to standardize the tracks for the Tiger B with the Henschel- and Porsche-Laufwerk.

The steel-tired, rubber-cushioned roadwheels themselves still present problems because all too easily the bearings are shoved axially and set crookedly so that the roadwheel discs tumble.

6.3.2 AUTOMOTIVE MODIFICATIONS

The following detailed listing of the important improvements and modifications to the **Tiger B Fahrgestell** were included in an attachment to the **Heerestechnisches Verordnungsblatt** (Army Technical Orders) from 15 December 1944:

6.3.2.1 Wechselgetriebe-Oel-Kuehler (Transmission Oil Cooler)

Starting with **Fgst.Nr. 280001**, a **Verdampfungskuehlanlage** (evaporator cooler system) was installed to lower the oil temperature in the transmission and prevent problems in shifting.

6.3.2.2 Kettenabdeckung (Track Guards)

Starting with **Fgst.Nr. 280001**, the **Kettenabdeckung** (track guards) were redesigned as curved sections with detachable side extensions for covering the wider cross-country tracks.

6.3.2.3 Maybach HL 230 P30 Motor

In November 1943, starting with **HL 230 P30** motor number 8322575, the governor was preset at the factory for a maximum speed of **2500 rpm** under full load and the motors were outfitted with a hand-operated temperature control on the oil cooler.

Maybach HL 230 P30 motors with serial numbers from 8321812 to 8322581 and motors that were rebuilt in October and November 1943 (with M stamped on the serial number plate) had faulty bearings that frequently failed. Improved bearings were installed in **HL 230 P30** motors starting with serial number 8323426 in January 1944.

Starting in March 1944, a new piston design was installed in the **Maybach HL 230 P30** motors (starting with serial number 8324290 at Maybach and 8322110 at Auto Union), reducing the compression ratio from 1:68 to 1:64.

On 7 March 1944, units were advised that some of the **Maybach HL 230 P30** motors had been outfitted with a **Durchdrehanlasser** (reduction gear hand-cranked starter) instead of a **Schwungkraftanlasser** (inertia hand-cranked starter).

6.3.2.4 Teleskoprohr (Telescoping Air-Intake Tube)

A telescoping tube for driving submerged up to a depth of 4 meters of water was installed in only a few **Tiger Ausf.B** produced up to March 1944. In its retracted position the telescoping tube was protected by the same hinged armor cap that had been designed for the **Panther Ausf.D**. When not installed, this air intake opening for the engine compartment was covered by a wire mesh screen held by a bolted circular flange. The support pipe for the telescoping tube was still installed underneath the rear deck, protecting the fuel filler tank and engine compartment from shell fragments.

The reason for dropping the requirement for submerged driving was provided in the report on the interrogation of Dr. Stiele von Heydekampf, President of the Panzer Kommission and managing director at Henschel, on 5 and 6 June 1945: Although submerging equipment was not made since early 1943 (50 Tiger I tanks were equipped), the demand for submerging ability for the Tiger II tank was not relaxed until mid year 1944. Requirements for wading to a depth of 180 cm were continued. The reason given for abandonment of total submersion was that they had expected that rivers would have to be waded but it was found that the standard 16-ton engineer bridge would carry the Tiger II.

6.3.2.5 Auspuffanlage (Engine Exhaust System)

To prevent hot exhaust gases from being sucked into the cooling system, the straight exhaust pipes with cover plates were replaced with bent pipes like those on the Panther. While **Fgst.Nr. V1** and **V2** had bent pipes, some of the production series **Tiger Ausf.B** completed in January through March 1944 still had straight exhaust pipes. Production series **Tiger Ausf.B** completed up to May 1944 also had a sheet metal heat guard surrounding both types of exhaust pipes.

6.3.2.6 Kuehlwasserheizgeraet (Coolant Heater)

Starting in February 1944, a **Kuehlwasserheizgeraet** was mounted on the left side of the **Maybach HL 230 P30** engine. An access port with an oval-shaped armor cover held by two fasteners was located in the tail plate below the armor guard for the left-hand exhaust pipe. To heat the engine coolant prior to attempting to start the engine, the armor cover was removed and a blow torch suspended from a hanger on the tail plate.

6.3.2.7 Hull Armor

Starting in April 1944, to improve the radio operator's field of view, a notch was cut out of the protruding edge of the hull glacis plate in front of his periscope.

The face of the hull side extensions at the front, drilled for mounting tow shackles, was cut out to allow additional room for fitting C-hooks. The hull side extensions at the rear, drilled for mounting tow shackles, were also reshaped to allow the tow hooks to pivot upward.

6.3.2.8 Kraftstoffanlage (Fuel System)

Starting in April 1944, fuel tank installation was altered and a vent line was added at the high point in the system.

6.3.2.9 Triebrad und Gleiskette (Drive Wheel and Track)

Starting in May 1944, the original **Gg 24/800/300** track was replaced by a new design **Gg 26/800/300** cross-country track with a single cast piece connecting link without holes for the sprocket teeth. Drive sprockets for the new two-piece track only had nine teeth instead of eighteen. Later the arrangement of the holes bolting the sprocket ring to the drive wheel were offset from the teeth as shown on drawing 021 B 49512 dated 6 August 1944.

6.3.2.10 Gelenkwellen (Drive Shafts)

Starting with **Fgst.Nr. 280063**, completed in late June 1944, a spring was strengthened to pre-tension of 45.6 kilograms to prevent damage to the teeth in the side drive shafts.

6.3.2.11 Kuehlung (Engine Cooling System)

Starting with **Fgst.Nr. 280067**, completed in late June 1944, the slip clutch in the cooling fan drive was modified to improve its functioning by increasing the size of the bearings and improving their seat.

6.3.2.12 Wechselgetriebe-Oel-Kuehler (Transmission Oil Cooler)

Starting with **Fgst.Nr. 280080**, completed about 9 July 1944, the thickness of the bottom of the cooling tank was increased from 1.5 mm to 5 mm and a rubber pad 480 mm x 40 mm x 5 mm was installed under the support.

6.3.2.13 Sammler-Isolierkaesten (Battery Isolation Boxes)

Starting with **Fgst.Nr. 280090**, completed about 16 July 1944, the isolation box for the battery was replaced by one with heated air connections for the upcoming **Kampfraumheizung** (fighting compartment heater).

6.3.2.14 Anschlagbock (Bump Stop)

Starting with **Fgst.Nr. 280127**, completed about 5 August 1944, installing and removing the bump stops for the roadwheel suspension arms was made easier by modifying the fastener so that the bolts could be tightened and loosened from the outside.

6.3.2.15 Luefterklappen (Ventilation Louvers)

The adjustable ventilation louvers in front of the radiators hadn't proven to be functional. The louvers and the control rods were no longer installed starting with **Fgst.Nr. 280150**, completed about 13 August 1944.

6.3.2.16 Kraftstoffbehaelter (Fuel Tank)

Starting with **Fgst.Nr.280161**, completed about 17 August 1944, in place of threaded stoppers, drain valves were built into the fuel tanks mounted on the side of the fighting compartment and both fuel tanks under the radiators.

6.3.2.17 Lenkbremse (Steering Brake)

The lubrication points for the steering brake linkage were not accessible. Starting with **Fgst.Nr.280210**, completed on 1 September 1944, openings were cut into the brake covers so that the lubrication points were accessible.

6.3.2.18 Kraftstoffleitung (Fuel Lines)

Connections for the fuel lines at fire wall penetrations were difficult to get at because they lay underneath a torsion bar. Starting with **Fgst.Nr. 280213**, completed on 2 September 1944, the penetration was changed by welding pieces of pipe to the fire wall so that the fuel line connection no longer lay underneath the torsion bar.

6.3.2.19 Oeffnung fuer Lufteintritt (Air Intake Opening for the Engine Compartment)

Starting with **Fgst.Nr. 280225**, completed on 10 September 1944, when the support pipe for the telescoping tube was no longer installed, the air intake opening for the engine compartment was protected from shell fragments by covering it with an armor cap that was bolted on.

6.3.2.20 Kraftstoffanlage (Fuel System)

The fuel tanks underneath the radiators had torn out of their fasteners. Starting with **Fgst.Nr. 280250**, completed about 15 September 1944, the bands holding the fuel tanks were welded at a second point.

By October 1944, the number of fuel line connections was decreased by about half to decrease fuel leaks in the engine compartment. The reduced number of interconnections between fuel tanks necessitated the addition of three vent lines to ventilate all seven fuel tanks.

6.3.2.21 Tool Stowage

Starting in October 1944, the brackets for holding a jack were no longer welded onto the tail plate below the engine access hatch because jacks were no longer being issued with the **Tiger Ausf.B**. The brackets for holding the wooden jack block and the fasteners for the track cable were both dropped by March 1945.

6.3.2.22 Huelsendeckelverriegelung (Hatch Cover Latch)

The latches for the driver's and radio operator's hatches tended to get stuck. Starting with **Fgst.Nr. 280302**, completed in November 1944, the original latch was replaced with two key locks that worked separately.

6.3.2.23 Kettenabdeckung (Track Guards)

Starting in March 1945, the curved track covers on both sides at the front were reinforced with vertical ribs.

6.3.2.24 Gleisketten (Tracks)

Starting in March 1945, a new single-link **Kgs 73/800/152** track was introduced that was again driven by an 18-tooth drive sprocket.

6.3.3 TURRET AND WEAPONS MODIFICATIONS

6.3.3.1 Turmpanzerung (Turret Armor)

Originally, Krupp completed all 50 turrets for the **VK 45.02 (P)** with 25 mm thick fore and aft roof plates. Only the center horizontal plate for mounting the cupola, loader's hatch, and ventilation fan was 40 mm thick. The three turrets mounted on the **Versuchs-Serie** chassis (**Fgst.Nr. V1**, **V2**, and **V3**) still had turret roofs with 25 mm thick plates fore and aft. This is the configuration of Tiger II (**Fgst.Nr. V2**) with **Turm Nr. 150110** at the Tank Museum in Bovington. However, **Turm Nr.1-50**, installed on the production series chassis starting with **Fgst.Nr.280001**, were modified by cutting off the original 25 mm roof plates and replacing them with 40 mm thick plates. These modified turrets were assigned new **Turm Nr.** in the 280001 series. The front roof plate was now flush with the top edge of the turret front plate, and there wasn't any 15 mm step up to the 40 mm center plate. The turrets for the production series were further modified by adding a square deflector bar which was welded across the front of the turret roof. In addition, a drain channel was cut into the top of the 40 mm roof plate to the right of the loader's hatch.

At least 16, but not all 50, of the original turrets fabricated for the **VK 45.02 (P)** had a hole cut in the left turret side for the commander's **Schaulock** (viewing port). As already decided in November 1942, a plug was welded into the hole intended for the **Schaulock**. Covered by Zimmerit on production series Tiger II, this plug was only discernible as a flatter circular area in the curved turret side.

6.3.3.2 Turmfugenschutzring (Turret Ring Guard)

Drawing number 021 B 49501 dated 14 January 1944 for the hull deck was labeled as applicable only to the **1. bis 50.Wanne** (1st through 50th hull) because it contained **Veraenderung f = Turmfugenschutzring** (change f for a turret ring armor guard). This armor ring (2420 mm inner diameter) was made up of arc segments, 100 mm thick at the base, 54 mm thick at the top, and 100 mm high. It was introduced on production series **Tiger Ausf.B** starting in April 1944.

6.3.3.3 Turmzielfernrohr 9d (Monocular Telescopic Sight)

On 2 February 1944, Oberstlt. Crohn (**Wa Pruef 6 IId**) informed Krupp that in a short time only monocular instead of binocular sights would be installed in Tiger turrets. He requested that Krupp propose a suitable shot-proof armor plug to reseal the left opening that had already been bored in the turrets.

Leitz reported completion and delivery of their first 10 **T.Z.F.9d** telescopic gun sights in April 1944. Starting in April 1944, monocular **T.Z.F.9d** gun sights replaced the binocular **T.Z.F.9b/1** gun sights in turrets completed by Wegmann.

6.3.3.4 8.8 cm Kw.K.43 mit geteilte Rohr (Sectional Gun Tube)

On 19 February 1943, Krupp proposed to **Wa Pruef 4** to convert the **8.8 cm Kw.K.43** series production to the **geteilte Rohr mit dem Laengsbeweger** (sectional gun tube with the longer sleeve) starting with number 200. Krupp stated that it could be done sooner at 100, but they didn't want to interrupt production. With the cancellation of the **VK 45.02 (P)** project there was no longer any rush and production was converted to assembling **8.8 cm Kw.K.43 L/71** with sectional gun tubes by March 1944. The **8.8 cm Kw.K.43 L/71 mit geteilte Rohr** was mounted in completed Tiger Ausf.B (**Fgst.Nr.280018**) before the end of April 1944, although others (**Fgst.Nr.280035** and **280037**) produced in late May and early June 1944 were outfitted with monobloc guns. Starting in June 1944, a lighter, smaller muzzle brake was fitted to the sectional guns. Originally, a replaceable, caliber diameter ring was threaded into this lighter muzzle brake, but this was discontinued in July 1944.

6.3.3.5 Zielstachel (Blade Sight)

Starting in May 1944, a **Zielstachel** (forward blade sight) was mounted on the left front corner of the turret roof. The **Kimme** (rear sight consisting of two wires) was welded to the armor cover for the periscope. The commander was to use this rough sight to determine when the gunner was aligned on selected targets instead of using the azimuth indicator ring in the cupola which the blade sight replaced.

6.3.3.6 Pilze fuer Behelfskran 2t (Sockets for Jib Boom)

Starting in June 1944, a **Behelfskran 2t** (jib boom) was issued to the troops to aid in tank repairs. This **Behelfskran** was mounted on three **Pilze** (sockets) welded to the turret roof. The drawing, showing the specific locations for welding three **Pilze** onto turret roof of the Tiger II, was distributed on 14 June 1944. The **Behelfskran 2t** could be used to lift the rear decking, motor, and transmission with steering gear from the Panzer on which it was mounted or to lift components from an adjacent vehicle.

6.3.3.7 Serienturm (Production Series Turret)

Starting with **Fgst.Nr.280048**, completed in June 1944, the **Serienturm** was mounted on the **Tiger Ausf.B Fahrgestell**. As described in manual D656/42 dated 1 July 1944 for the **Panzerkampfwagen Tiger, Ausfuehrung B, Turm Serie**, there were many more changes introduced with the **"Serienturm"** than just a thicker front plate, elimination of the bulge on the left side plate, and a redesigned gun mantlet, as follows:

The turret front plate is set at an angle of 10° from vertical. The side walls are the same as the rear wall at 20°. The turret roof consists of a forward plate set at 78°, a middle horizontal plate, and a rear plate set at 82°.

*The turrets for **Tiger Ausf.B Nr.1-50** and the **Serienturm** at and above **Nr.51** are not interchangeable on the Tiger II chassis without making additional changes. The **Serienturm** (at and above **Nr.51**) can be mounted on **Fahrzeuge Nr.1-50** (**Fgst.Nr.V1**, **V2**, **V3**, and **280001** through **280047**) by complying with the following points: 1) bore out the 24 holes for the fastening bolts from 24 mm to 27 mm, 2) position 5 mm thick strips on the flange located on the chassis deck in order to compensate for the difference in the turret clearance, and 3) remove any **Fugen-Schutzsegmente** (turret ring guard segments) from the deck of the chassis.*

***Turm Nr.1-50** can be mounted on **Fahrzeuge ab Nr.51** (**Fgst.Nr.** at or above **280048**) if 1) the turret seal for submerged fording and the turret alignment studs are removed and 2) shims are positioned under the turret fastening bolts to compensate for the holes that are too large.*

*The **8.8 cm Kw.K.43** gun tube can be removed from the **Serienturm** though the hatch in the turret rear without removing the hatch cover. But both rear ammunition racks that protrude into the hatch profile have to be unbolted. The rear hatch cover has to be dismounted in order to remove the **8.8 cm Kw.K.43** gun carriage from the turret. As in removal of the gun tube, the ammunition racks can remain in place. The carriage itself must be disassembled into the carriage body and the deflector guard. The disassembled ammunition racks can be removed from the turret without removing the rear hatch cover.*

*In **Turm Nr.1-50**, the turret rear wall has to be unbolted in order to remove the ammunition rack, the gun tube, or the gun carriage. However, the assembled component parts can fit through the opening. The trunnions must be unscrewed from the gun carriage before it is removed; this can't be avoided because of the space situation.*

*The **Turmkugellager** (turret ring ball bearing race) consists of two steel rings with semi-circular grooves in which 113 support bearings of 40 mm diameter and 113 **Trennringe** (separator rings) of 55 mm diameter run. The 24 bolts holding the inner race to the turret are outfitted with **Stauchringen** (shock rings), to take up shock if the turret is hit by a shell and to prevent the bolts from tearing out.*

Instead of an inflatable sealing tube, labyrinth seals are built into the turret ring. The grooves in the rings must be filled with grease to increase the effectiveness of the seal.

*The **Wiegenpanzer** (gun mantlet) covers the opening for the machinegun in the turret front plate, and it itself has a boring for firing bursts from the machinegun.*

The rear side of the machinegun aperture in the turret front plate is sealed against rainwater.

*The monocular **Turmzeilfernrohr 9d** is fastened at the front on the slides of the gunsight mount and at the rear by a pivoting arm suspended from the turret roof.*

*The two-part **Prismenspiegel** (loader's periscope) is encapsuled in a **Bakelit** housing. It is held in the periscope mounting by a bail, two bolts, a latch, cylinder, coil spring, and grip. A lip ring in the periscope mounting seals the closure of the periscope watertight.*

*The cast-armor ring of the **Kommandantenkuppel** (commander's cupola) is welded to the turret roof and secured with segment pieces. A **Zielstachel** (forward blade sight) is bolted to the turret roof in front of the forward-facing periscope. The **Kimme** (rear sight consisting of two wires) is welded to the armor cover for the periscope. With this device, the Panzer commander can recognize the aiming direction of the turret-mounted weapons and direct the gunner onto the selected target.*

*The rear turret wall is welded in place instead of bolted. The **Turmlukendeckel, hinterer** (rear hatch cover) has been redesigned but still has a **Rutschblech** (sheet metal inner cover), a **Drahtseil** (cable to pull the hatch closed), and an **MP-Stopfen** (pistol port with a plug).*

*The **Geschuetzmunitionslagerung** (ammunition racks) in the turret hold eleven **8.8 cm Patronen** on both the right and left sides. The ammunition racks were redesigned so that only the outer row is held by quick-release bands and the inner rows are held by spring-loaded wooden spacers.*

Two lights and a reading light are mounted under the turret roof. The amount of light can be adjusted by rotating covers which turn off the light when they are completely closed. In addition, an electric outlet is available for a hand-held light.

6.3.3.8 Einheitshalter fuer Scherenfernrohr (Standard Mount for Scissors Periscope)

On 4 May 1944, **Wa J Rue (WuG 6)** informed Henschel, Wegmann, and Krupp:

*Daimler-Benz has completed the final design for the **Einheitshalter fuer Scherenfernrohr** for **Pz.Kpfw. und Befehlswagen Tiger I und Tiger II**. The head was changed because it was only to be used for mounting the **Scherenfernrohr** instead of both the **Sehstab** and the **Scherenfernrohr**. The **Einheitshalter** was to be welded underneath the turret roof between the forward periscope and next periscope to the right. The holder for stowing the **Scherenfernrohr** when not in use was to welded under the turret roof to the left front of the Panzer commander on the outer bridge on which the junction box for the **Bordsprechgeraet** (intercom) was located. The **Einheitshalter** was to be available shortly from the firm Collignon, Berlin which was involved in its design.*

6.3.3.9 Ersatzkettenglieder (Spare Track Links)

On the subject of storing **Ersatzkettenglieder** on the **Tiger II Turm**, on 8 May 1944 Oberstlt. Crohn (**Wa Pruef 6/Pz IIb**) informed Henschel, Fried.Krupp, Rheinmetall-Borsig, and M.A.N. that:

The latest widely based test firing to investigate the effect on the protection against armor-piercing projectiles afforded by ar-

mor plate when covered by track links have provided the following results:

1. Protection is diminished when the armor plate is vertical or at angles up to 10° from vertical.

2. Protection is increased when the armor plate is at angles greater than 30° from vertical.

3. Protection remains unchanged when the armor plate is set at 10° to 30° from vertical.

*Therefore, there is no objection if **Ersatzkettenglieder** are mounted on the rear part of the turret side wall and the sloped walls of the hull.*

Starting in late June 1944, hangers were welded to the turret sides for storing two pairs of track links both fore and aft and on both sides of the turret.

6.3.3.10 Turmlukendeckel, Oberer (Loader's Hatch Cover)

Starting in late June 1944, a 40 mm thick **Turmlukendeckel** (loader's hatch cover) replaced the 15 mm thick forged cover. The hatch opening was cut straight without a rim, greatly reducing machining time.

6.3.3.11 Losterkennungstafel (Poison Gas Identification Panels)

At a meeting on 22 and 23 June 1944, Wegmann and **Wa Pruef 6** discussed the additional poison gas defense measures for the Tiger B. It was only possible to mount the **Losterkennungstafel** on the turret. Locations selected on the Tiger B were the first **Tafel** on the gun mantlet at an angle of about 15° and the second and third right and left on the rear of the turret deck at an angle of 25°. Each **Tafel** was secured on its base with two bolts. Wegmann stated that the modification would be implemented in the production series starting with the 98th turret, completed about 9 July 1944.

6.3.3.12 Kommandantenkuppel (Commander's Cupola)

By mid-July 1944, a redesigned periscope and periscope holder were introduced with a pivoting bail, two bolts to tighten the periscope flange against the seal, and a quick-release, spring-loaded latch. Starting in August 1944, a redesigned **Kommandantenkuppel** was mounted in a machined recess in the turret roof and fastened by seven bolts tightened against seven arc segments on the underside of the turret roof.

6.3.3.13 Armor Guards Over Torsion Bars

Starting in August 1944, a half-cylinder armor guard was welded onto each side of the lower turret rear as protection for the torsion bars that counterbalanced the rear hatch.

6.3.3.14 Regenschild (Rain Guard)

In September 1944, the armor manufacturers were ordered to initiate a modification by welding an inverted U-shaped guard over the gunsight aperture. This guard was designed to prevent rain from fouling the gun sight. Extending well beyond the turret front plate, the guard also reduced the angle at which the gunner would be blinded when trying to aim in the direction of the rising or setting sun. Due to the backlog of turrets already delivered by the armor manufacturers, this modification did not routinely appear until January 1945.

6.3.3.15 Schablonen (Stencils)

On 16 January 1945, Wegmann was informed that **OKH** had ordered that all writing be eliminated from inside the Tiger turret with the exception of the following:

1. "**lose - fest**" on the turret traverse lock
2. "**55 +5 kg/cm^2 braun ark**" on the recuperator
3. "**..... kg/cm^2 bei 0° Erhoehung**" on the pneumatic cylinder for the gun counter balance.

6.3.3.16 Fliegerbeschussgeraet (Anti-Aircraft MG Mount)

Starting in March 1945, the ring base welded onto the periscope guards was dropped and a single post (80 mm high, 25 mm diameter) was welded to the left front periscope guard to serve as the base for swiveling a new anti-aircraft machinegun mount.

6.3.3.17 Ersatzkettenglieder (Spare Track Links)

Starting in March 1945, three sets of track hangers were welded fore and aft on both sides of the turret for carrying spare **Kgs 73/800/152** track links.

6.3.3.18 Tarnung (Camouflage Rings)

Starting in March 1945, five semi-circular loops were welded onto both turret sides for use in attaching camouflage such as tree branches.

ABOVE AND LEFT: One of the first (if not the first) production series **Tiger Ausf.B** completed by Henschel in January 1944. The **Zimmerit** (anti-magnetic coating) hides the welded plugs for the **MP-Stopfen** (pistol ports) and **Schauloch** (view port) on the turret sides. The locations and details of the tool stowage, armor pot covering the telescoping air intake, and countersunk base for the radio antenna are discernible in these photos. The upper edge of the hull rear and back of the rear deck have been blackened by engine exhaust due to the baffle plates on the exhaust pipes. (TTM)

One of the first production series **Tiger Ausf.B** still in the Henschel assembly shop. It has been outfitted with the **Gg 24/800/300** cross-country tracks, curved front mudguards, hinged armor pot over the telescoping air intake, and exhaust pipes with baffle plates. (WJS)

Tiger Ausf.B (**Fgst.Nr.280006**), completed by Henschel in February 1944, was issued to the **Panzerversuchsstelle** in Kummersdorf. It is identified by "**211**" inside a circle and "**V6**" stenciled on the glacis plate. It was outfitted with the monobloc **8.8 cm Kw.K.43 L/71** gun with the heavy muzzle brake and the binocular **T.Z.F.9b/1** gun sight. It does not have a **Turmfugenschutzring** (turret ring guard) bolted to the top of the hull. (BA)

ABOVE: **Tiger Ausf.B** (**Fgst.Nr.280008**), completed by Henschel in February 1944, was issued to the **Panzerversuchsstelle** in Kummersdorf, who identified it by stenciling "**212**" in a circle and "**V8**" on the glacis plate. It has **Zimmerit** (anti-magnetic coating), the curved exhaust pipes, and the **Gg 24/800/300** cross-country tracks. (BA)

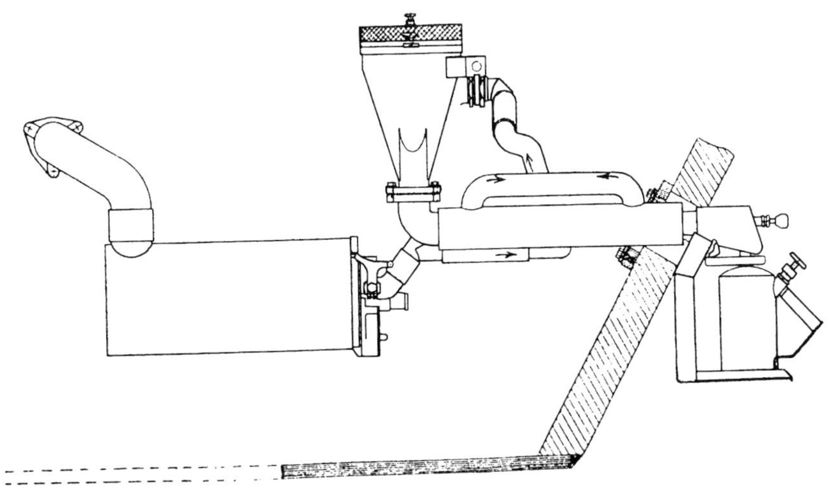

ABOVE AND RIGHT: Starting in February 1944, a **Kuehlwasserheizgeraet** (engine coolant heater) was installed. Flames from a blow torch heated the engine coolant in a cylindrical chamber around the central liner. Exhaust from the open flame was vented into the engine compartment.

THIS PAGE AND OPPOSITE: **Tiger Ausf.B** (**Fgst.Nr.280009** or **280012**), completed by Henschel in March 1944, was issued to **Wa Pruef 6** and retained at the Henschel testing area at Haustenbeck. It was used for a series of experiments, including testing the central poison gas filter and test-firing the gun. **Tiger Ausf.B** produced in March 1944 were still outfitted with the **Gg 24/800/300** cross-country tracks and did not have the **Turmfugenschutzring** (turret ring guard) bolted to the top of the hull. (WJS)

THIS PAGE AND OPPOSITE: These two **Tiger Ausf.B**, completed by Henschel in April and May 1944, like most issued to the **1.Kompanie/ schwere Panzer-Abteilung 503,** featured sectional **8.8 cm Kw.K.43 L/71** guns with the heavier muzzle brake, monocular **T.Z.F.9d** gun sights, drive sprockets with nine teeth for the double link **Gg 26/800/300** cross-country track with its single connecting link. The closeup shows details of the removable turret rear plate with the armor guard for the torsion bar which counterbalanced the weight of the rear hatch. (3-TTM, 1-NA)

THIS PAGE AND OPPOSITE ABOVE: **Tiger Ausf.B** (**Fgst.Nr.280031**), completed at Henschel in May 1944, was outfitted with the monocular **T.Z.F.9d** gun sight but still had a monobloc **8.8 cm Kw.K.43 L/71** gun. The turret, displaced by an internal explosion, was sent back to England for testing the armor quality (See Appendix D). Details can be seen of the 12-segment **Turmfugenschutzring** (turret ring guard) as well as wire mesh guard over the front cooling air intake grating for the radiators. (TTM)

Panzerkampfwagen Tiger Ausf.B - Fgst.Nr.280047 completed in June 1944 – **Pilze** sockets welded to turret roof – blade sight for commander – monocular **T.Z.F.9d** telescopic sight – sectional gun tube – turret ring guard – top edge of glacis notched to improve radio operator's view – modified hull side extension tow eyes (fore and aft) – double link **Gg 26/800/300** tracks with 9 teeth sprocket – curved tail pipes without sheet metal guard – engine coolant heater – no armor pot for telescoping air intake – vent on rear deck for motor compartment – vent line for fuel system.

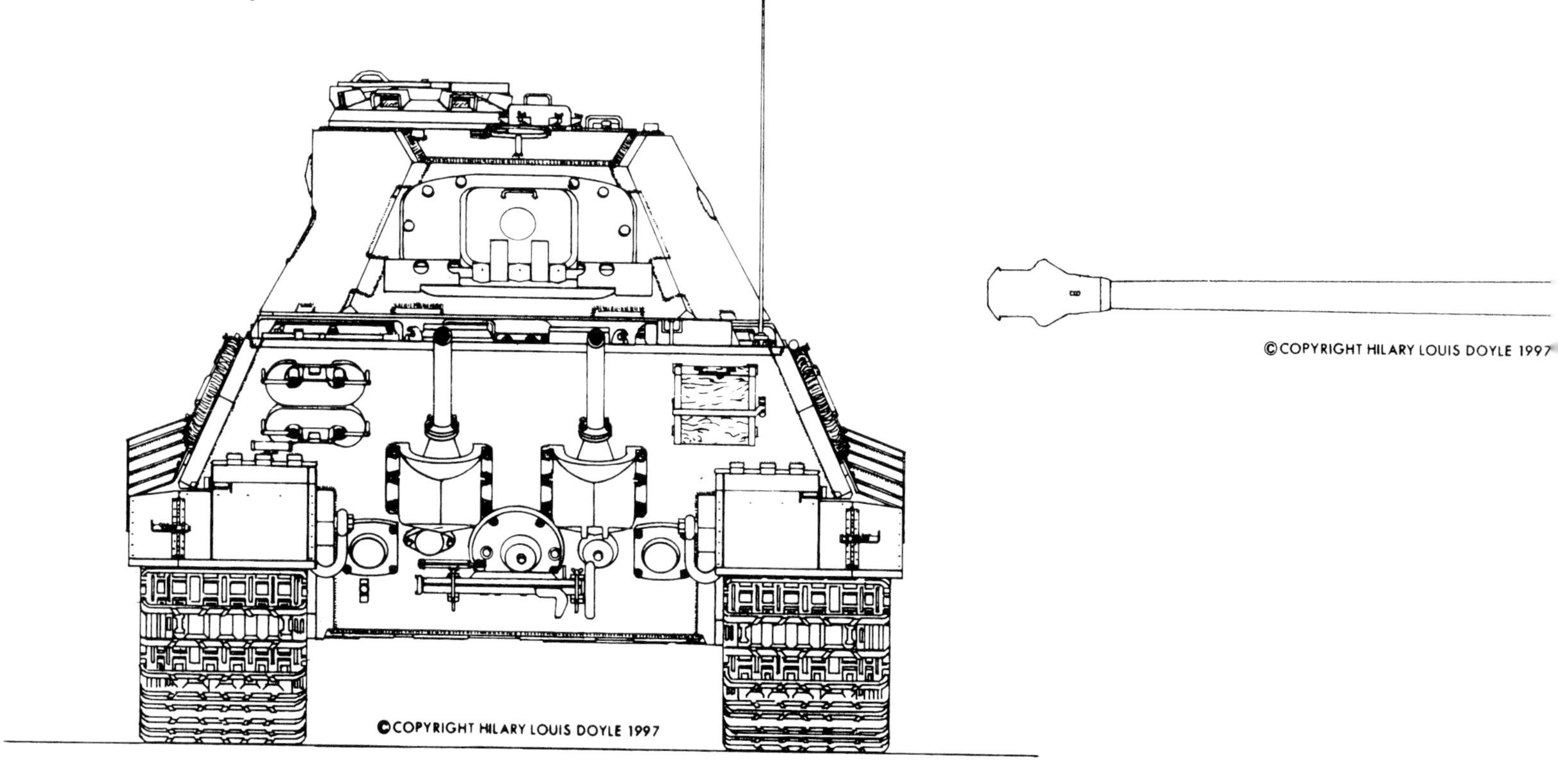

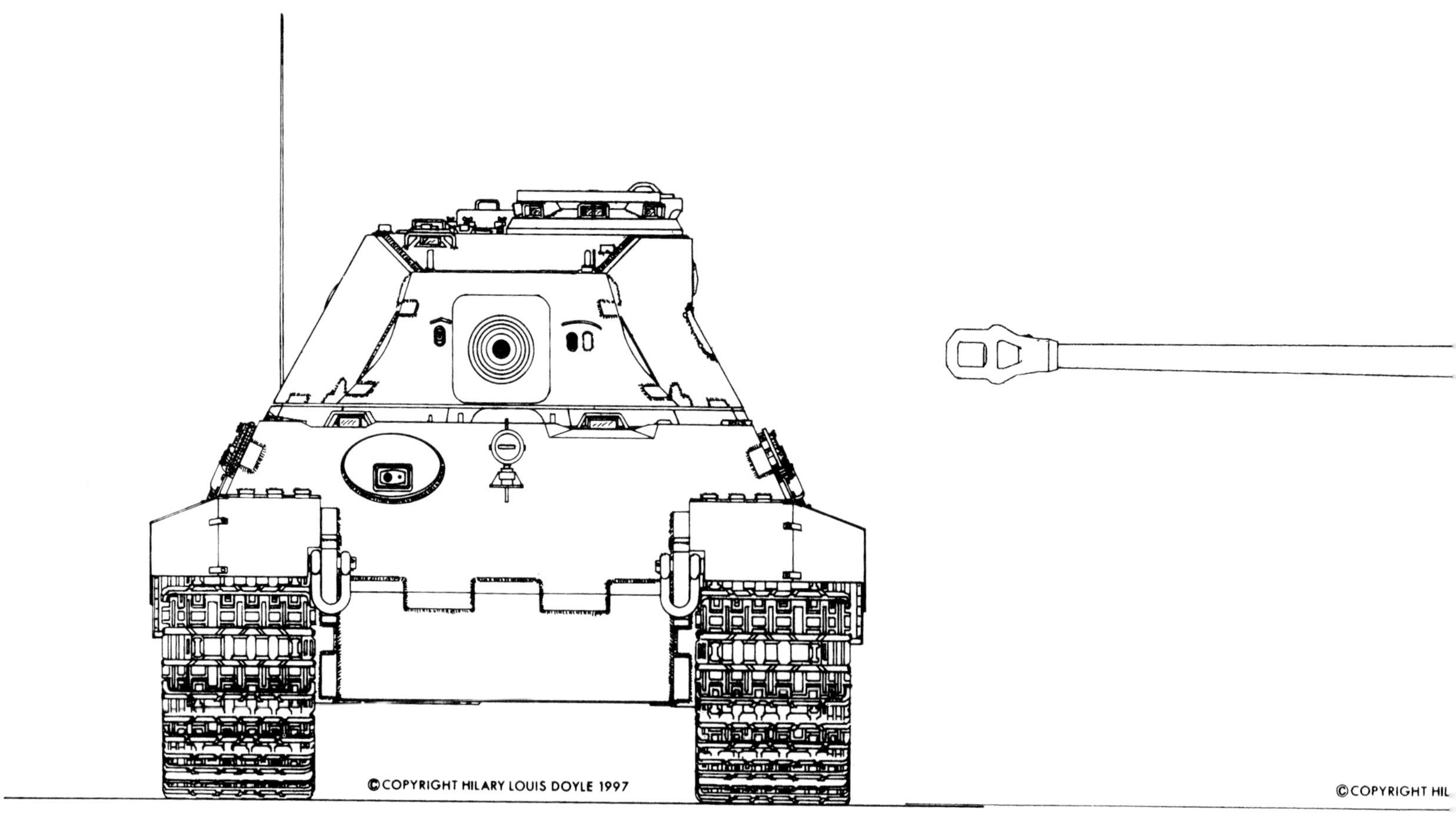

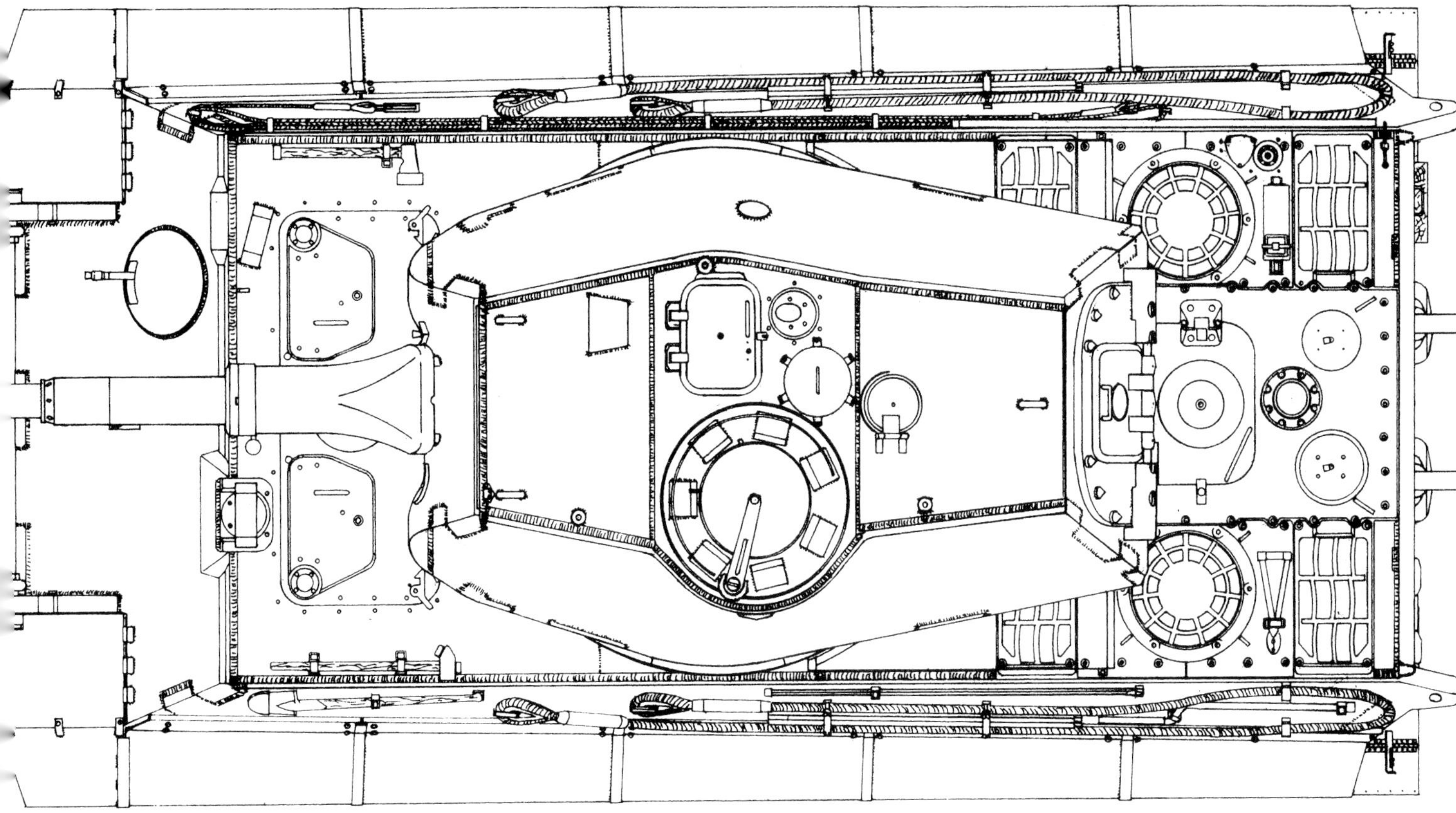

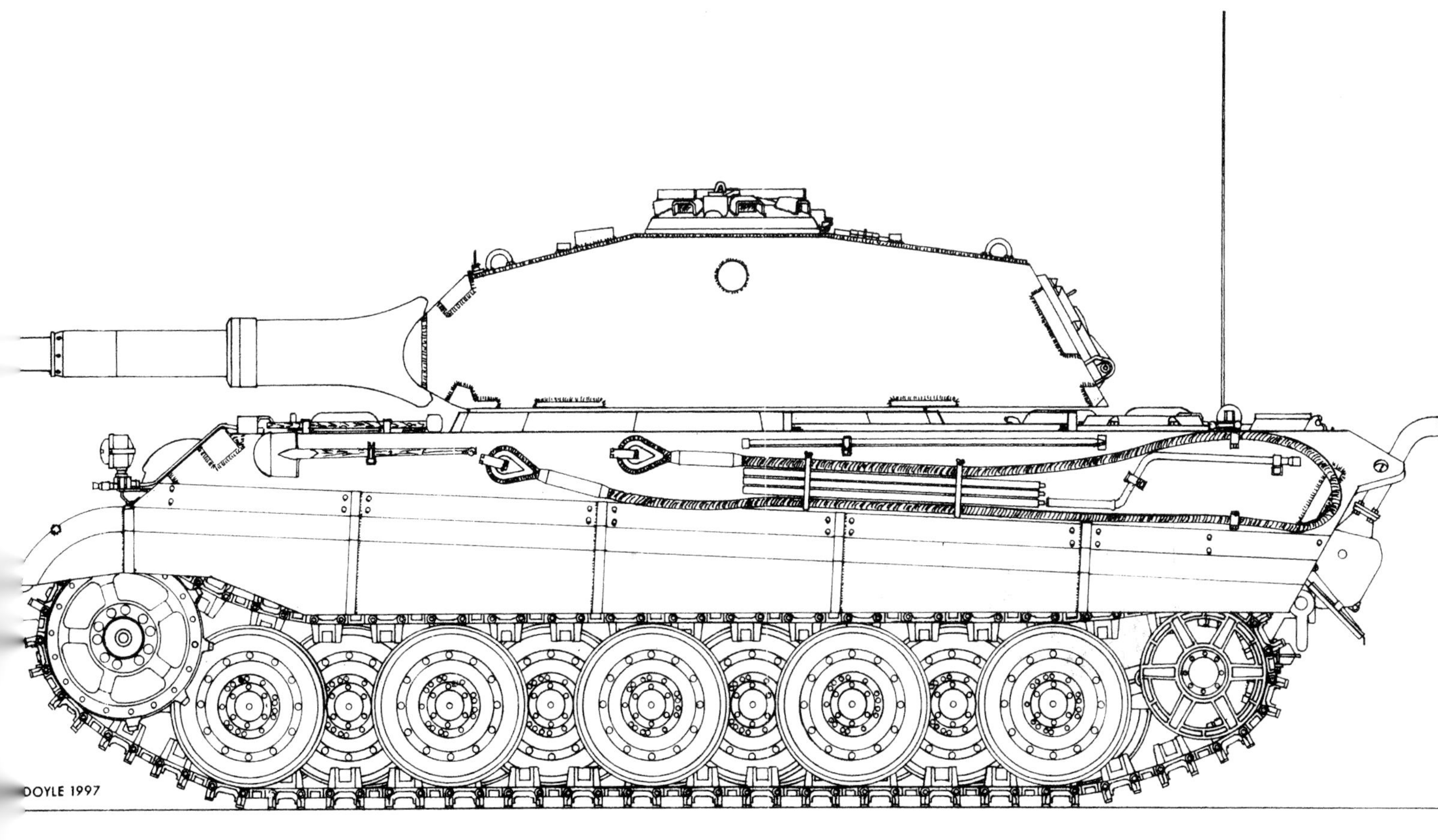
DOYLE 1997

Panzerkampfwagen Tiger Ausf.B - Fgst.Nr.280048 completed in June 1944 – **Serien-Turm** – 15 mm thick loader's hatch – commander's cupola welded to turret roof.

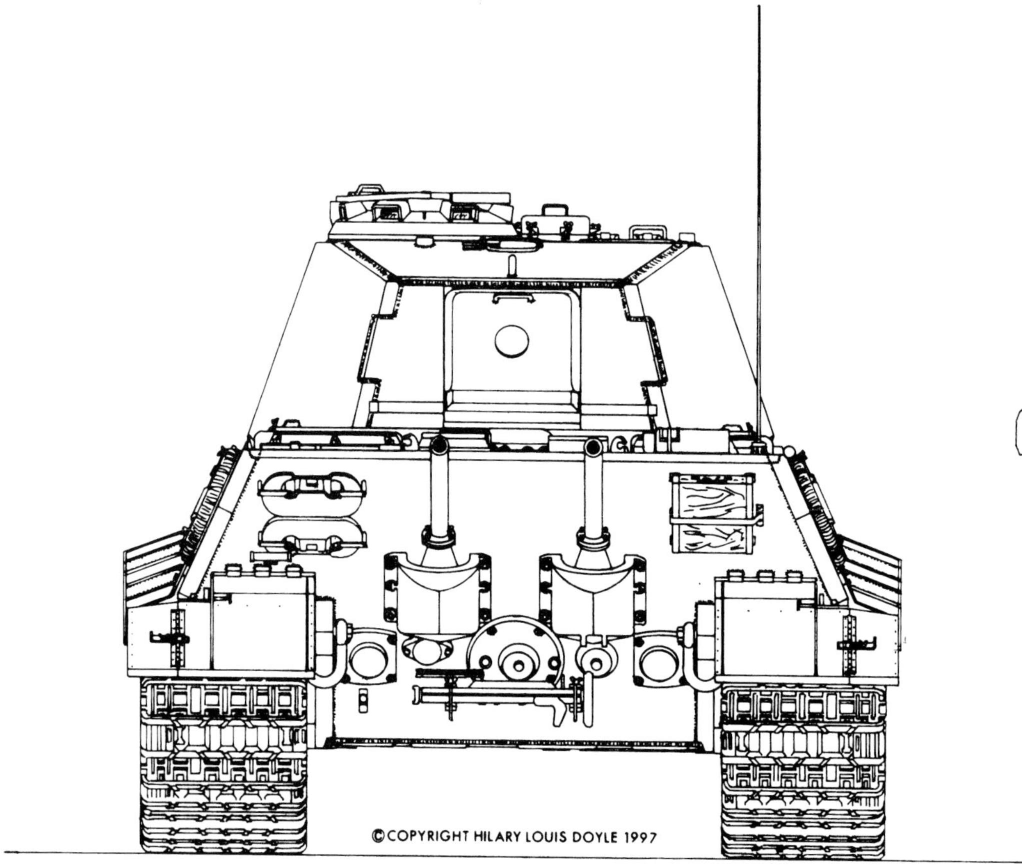

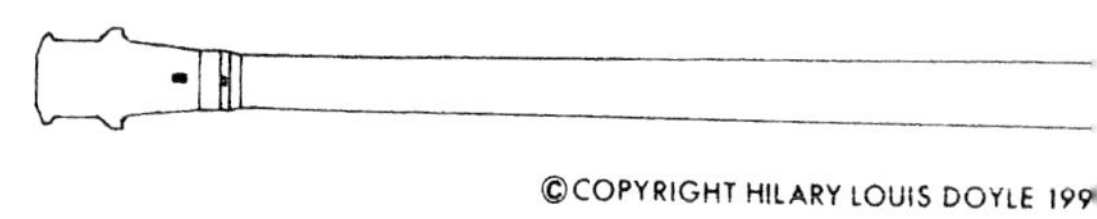

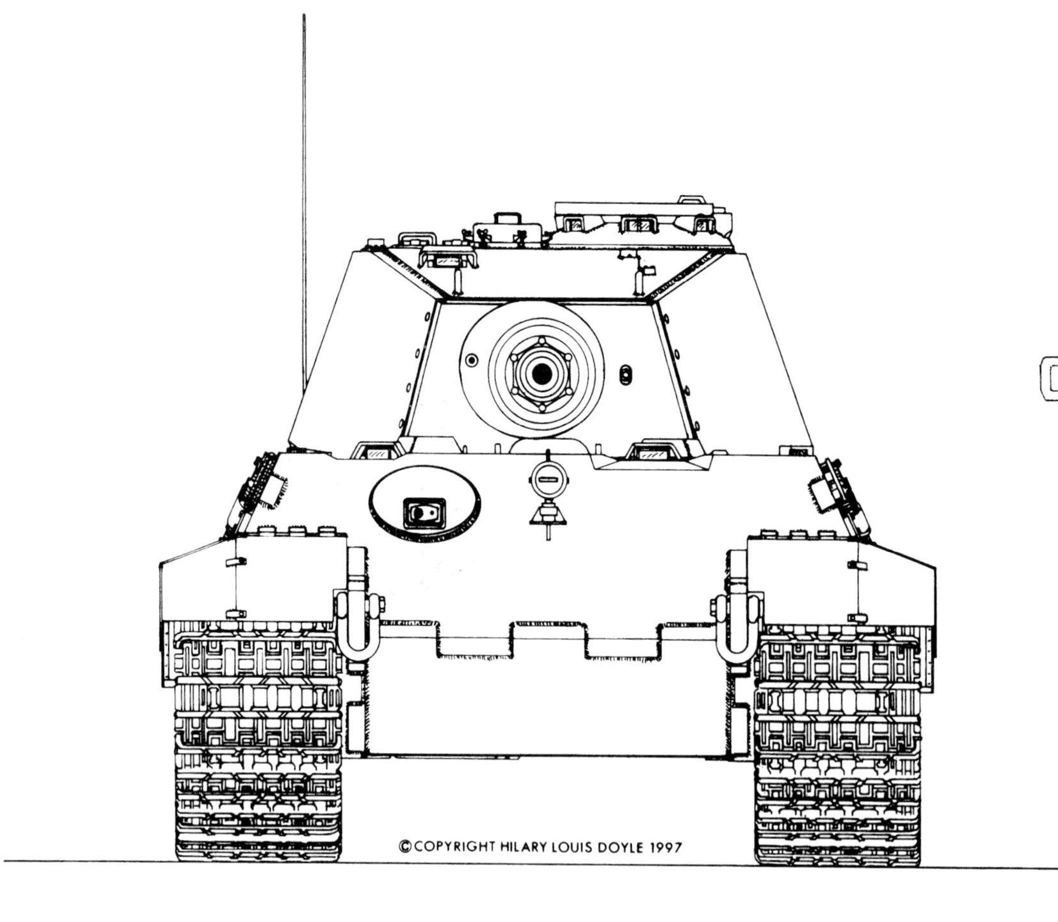

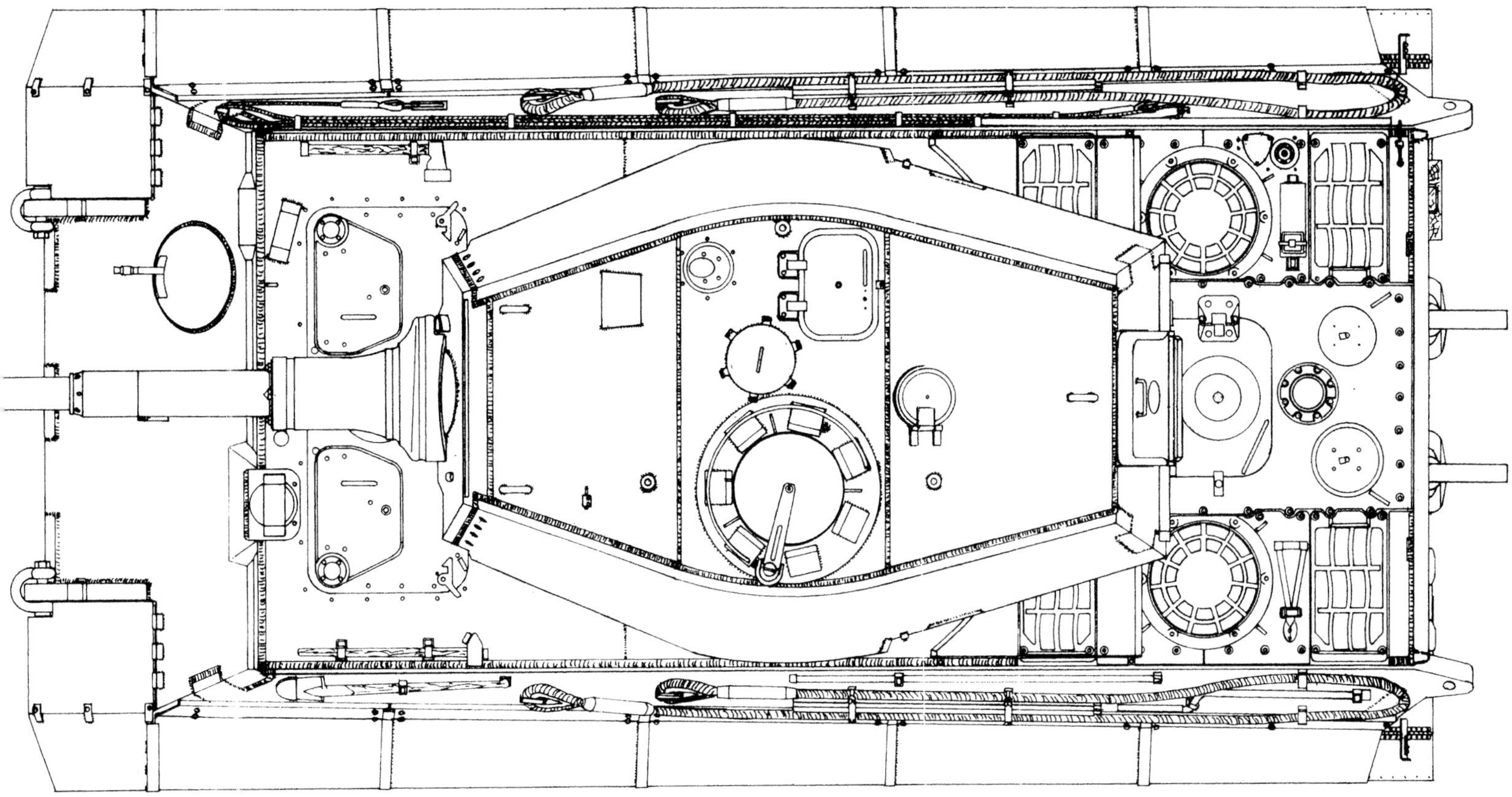

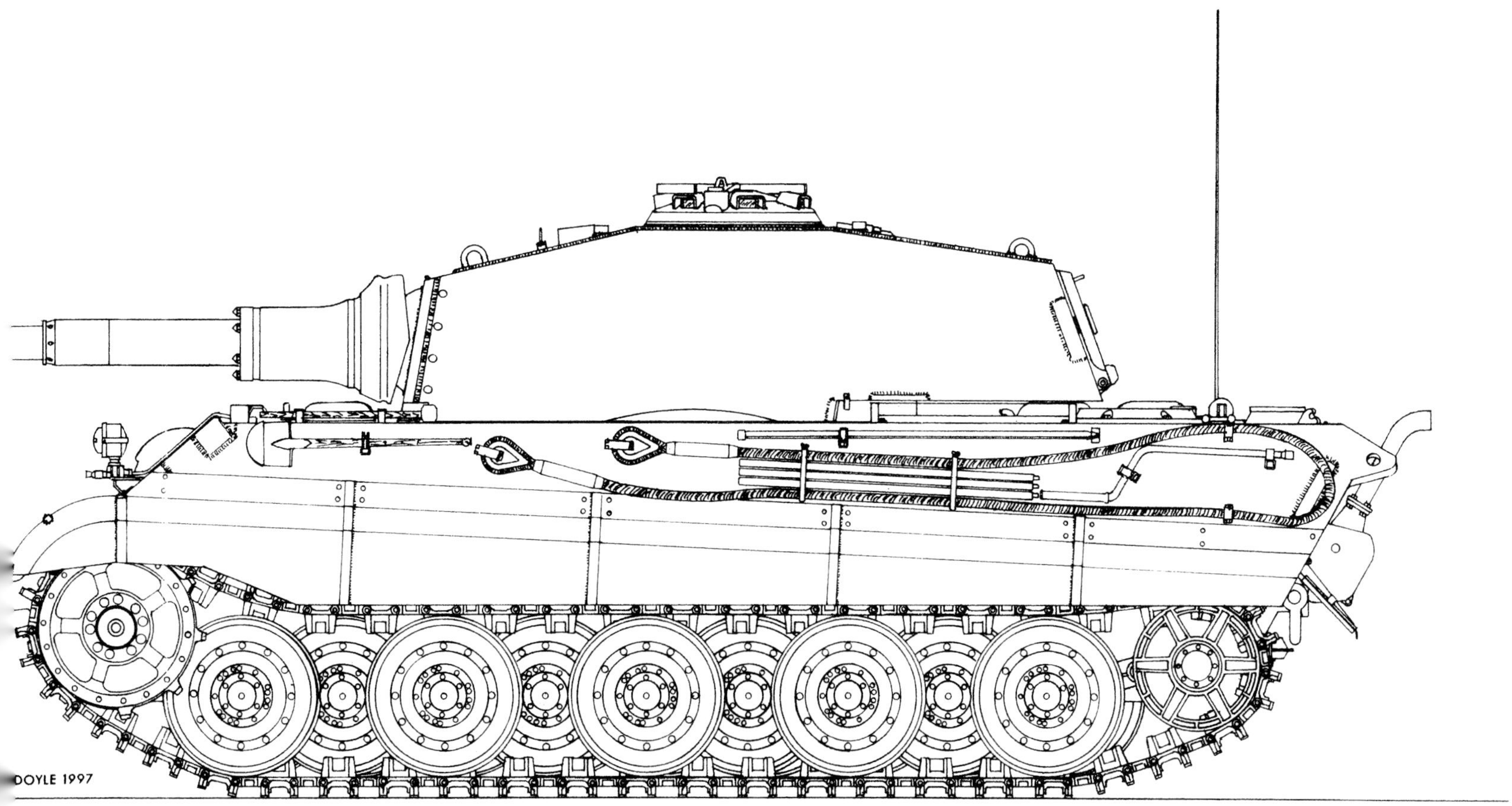
DOYLE 1997

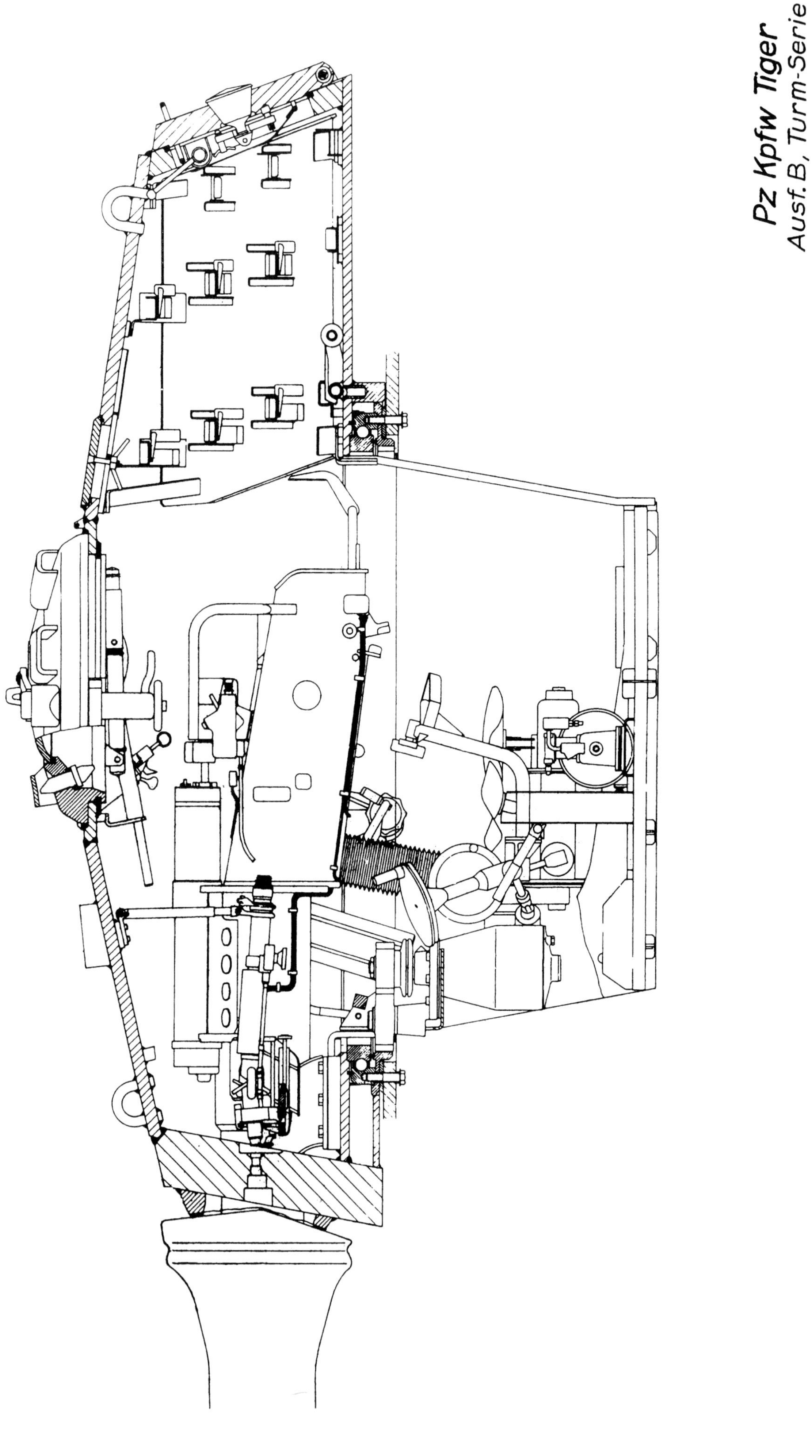

Arrangement of components inside the **Serienturm (021 B 50600 Bl.1)**.

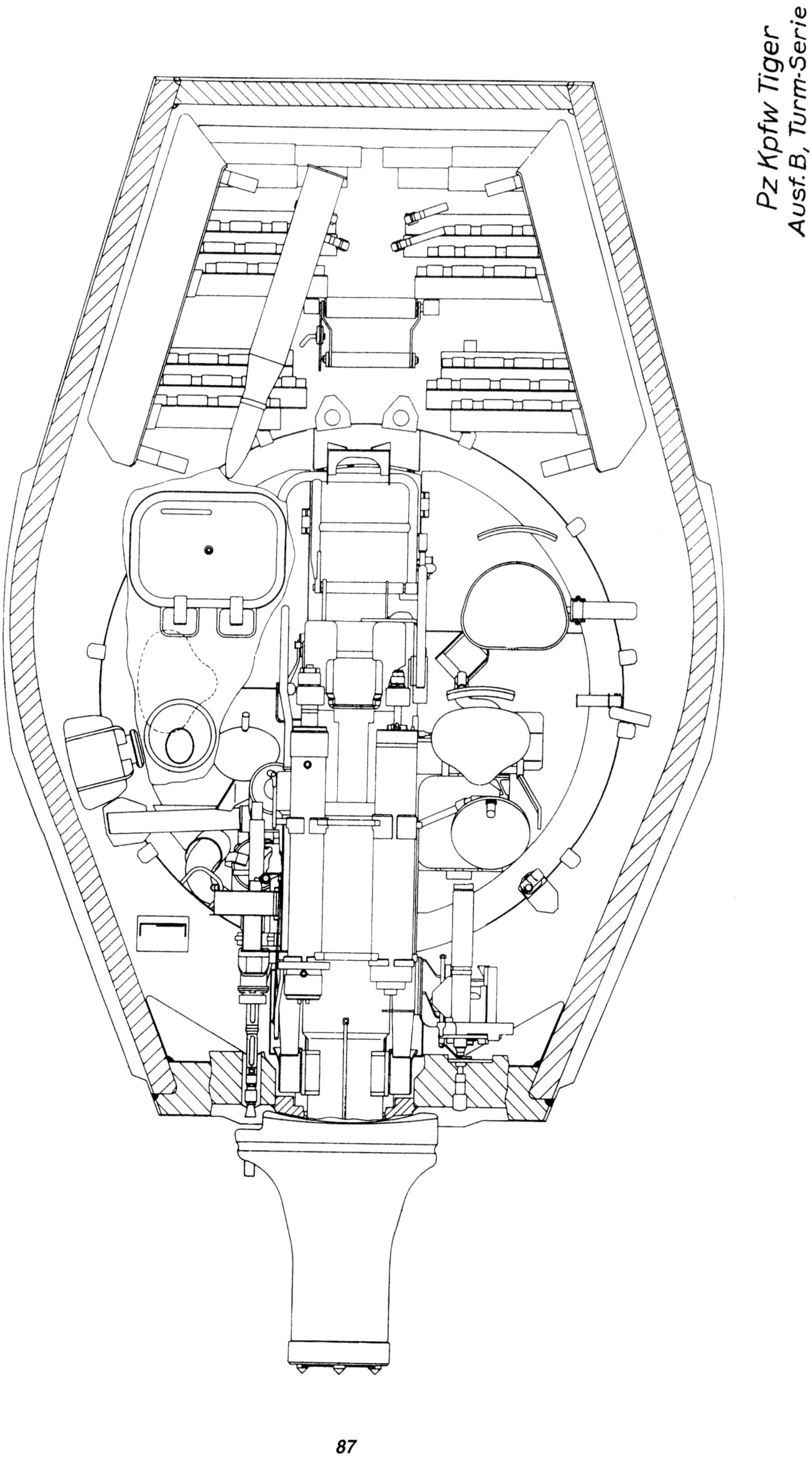

Arrangement of components inside the **Serienturm (021 B 50600 Bl.2)**.

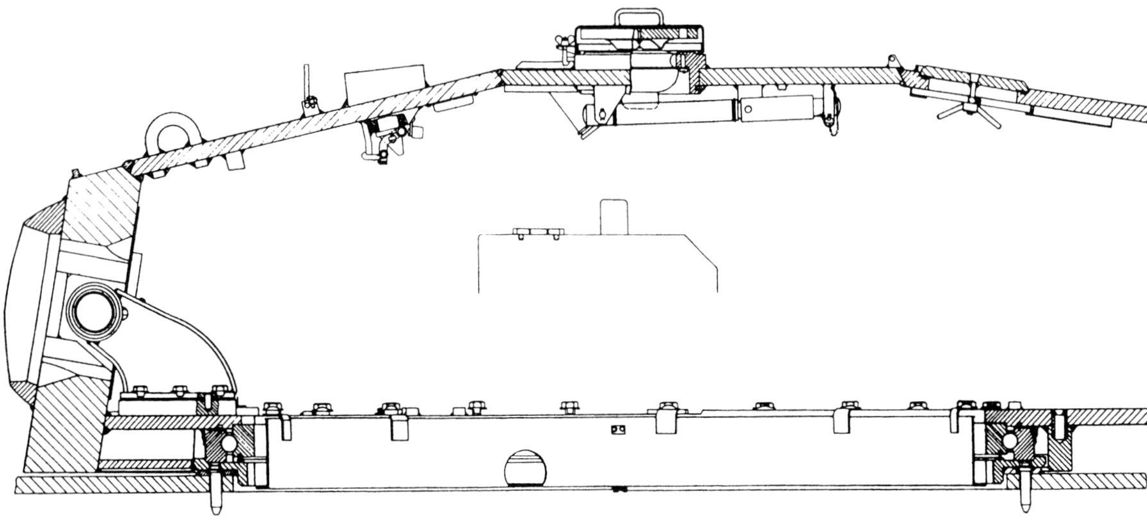

Turmgehaeuse (021 B 50601) turret armor body for the **Serienturm** with the interior travel lock mounted under the turret roof. The outline in the center is the location of the gun breach.

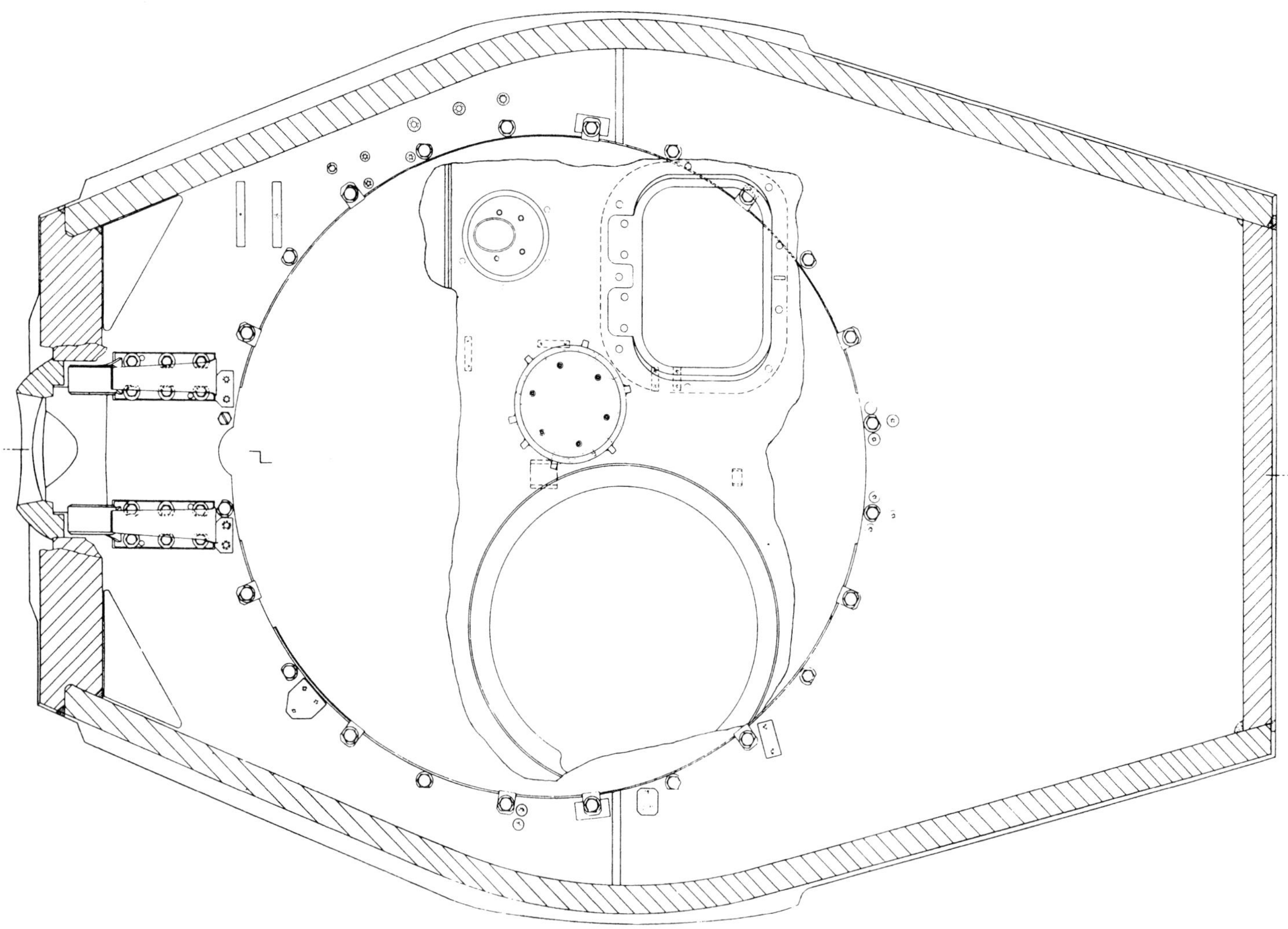

Turmgehaeuse (021 B 50601) turret armor body for the **Serienturm** with marks showing that the centerline of the gun was offset 30 mm to the right of the turret centerline. The turret's shape was asymmetrical. The left side was bowed 20 mm farther out than the right side.

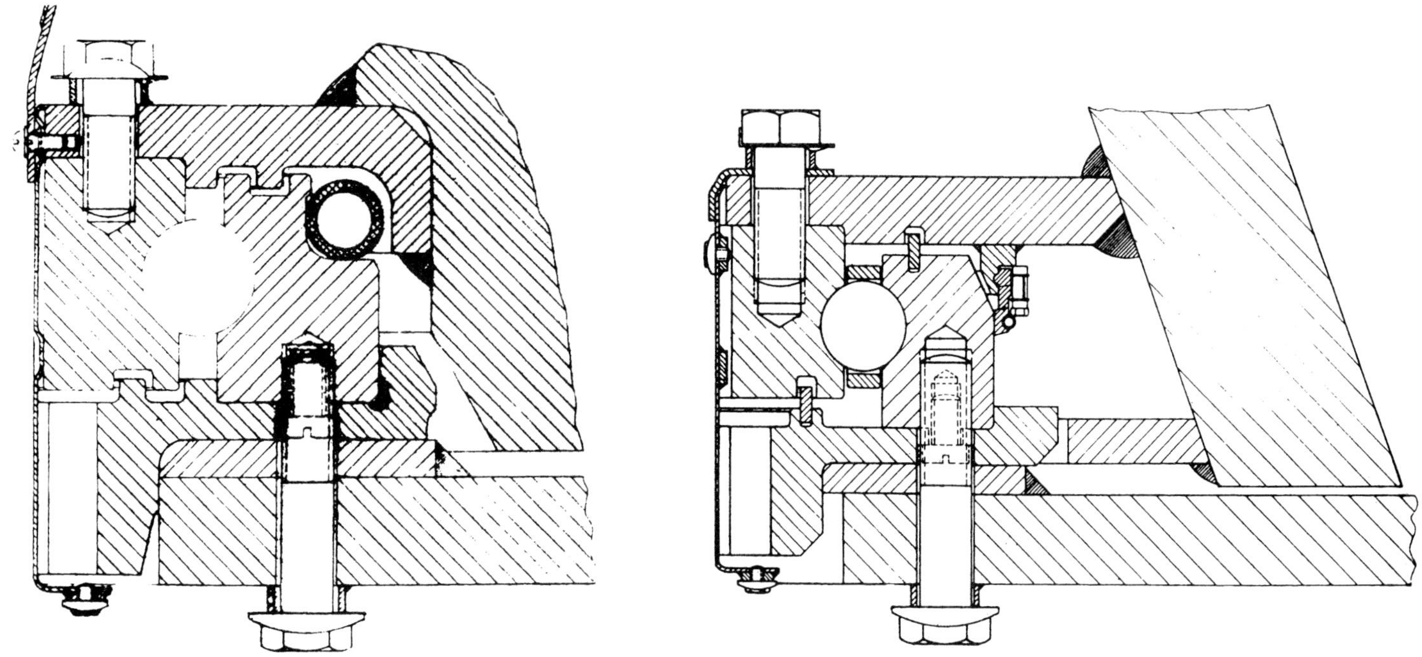

Turret ring for **Turm Nr.1-50** with an inflatable rubber sealing tube, double baffle seal, and ball bearings of two different sizes. As shown in the drawing on the right, the turret ring was redesigned for the **Serienturm** with ball bearings surrounded by spacer rings, simplified labyrinth seals, and larger diameter bolts.

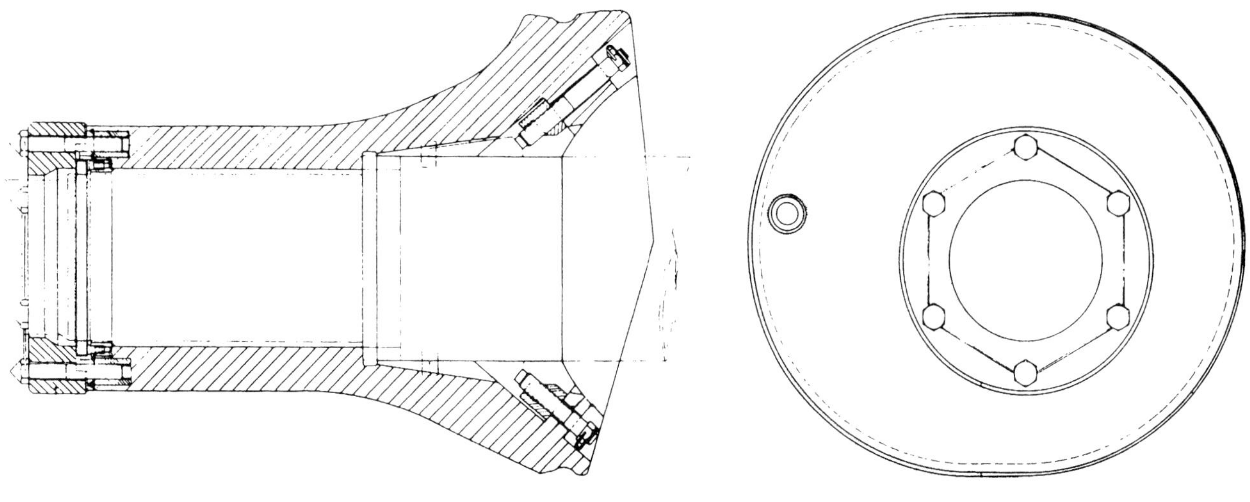

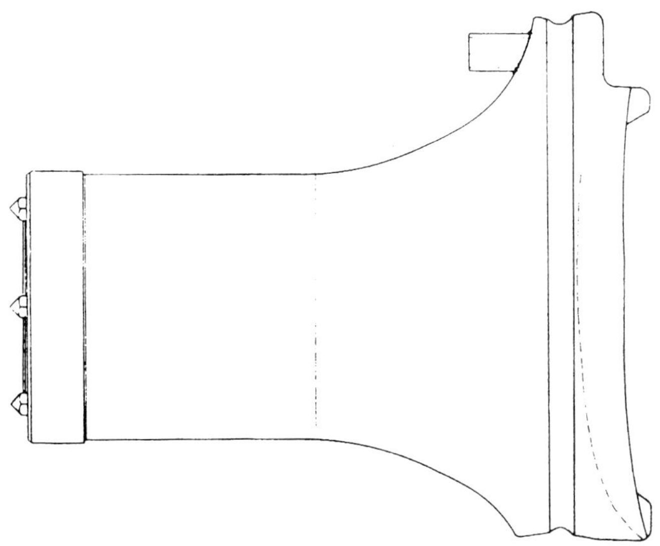

Wiegenpanzer (021 B 50604) gun shield, sectional views to show the way it was mounted and the seal to prevent accumulation of dirt inside the slide. The right side of the gun shield was extended to cover the aperture for the machinegun in the turret front. While present on the original drawing, the cylinder, welded to the hole through which the machinegun fired, has not appeared in photographs or on surviving Tigers.

The original **Pz-Fuehrerkuppel (021 B 2762)** commander's cupola for the **Serienturm** was welded to the turret roof. The design was adopted from the Tiger I. This drawing from the Tiger II manual still shows the 25 mm roof for the Tiger I. To mount it on a **Serienturm**, an additional 15 mm was cut out of the bottom of the armor ring so that it fit on the 40 mm thick turret roof. The two-piece periscopes were held in place by wing nuts. In addition, three channels were cut into the rim to drain rainwater away from the closed hatch cover.

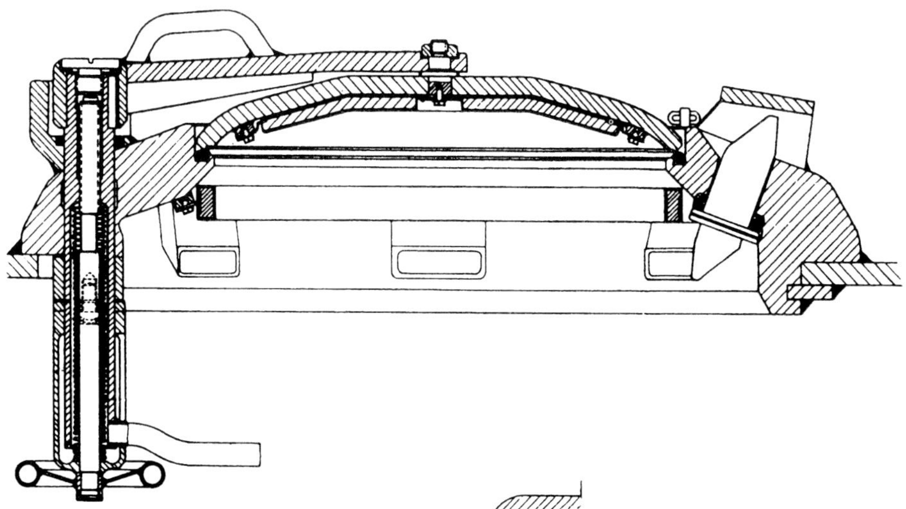

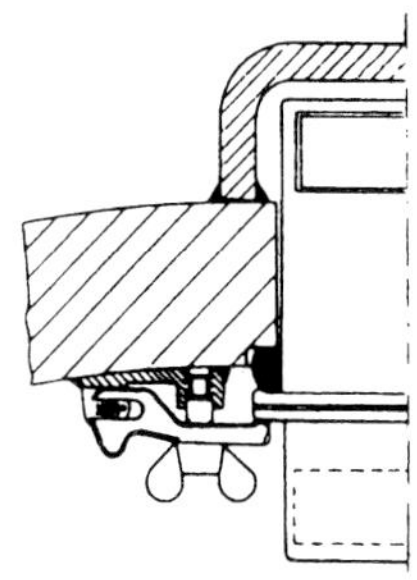

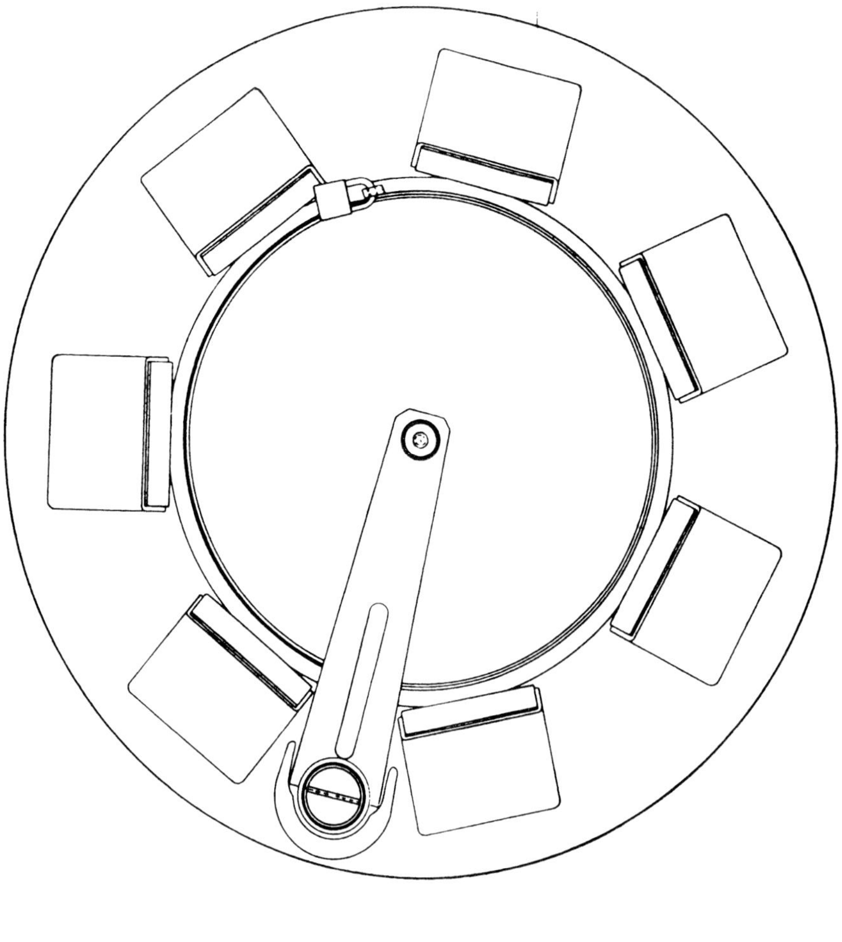

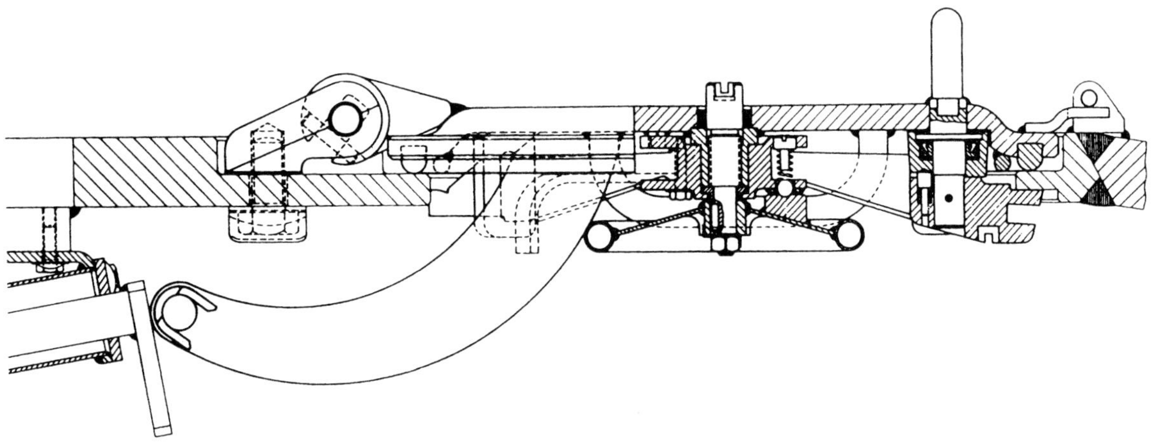

ABOVE AND BELOW LEFT: The 15 mm thick **Turmlukendeckel, oberer (021 B 50606)** (loader's hatch cover), introduced with the **Serienturm**, had shorter hinges than the loader's hatch cover on **Turm Nr.1-50**.

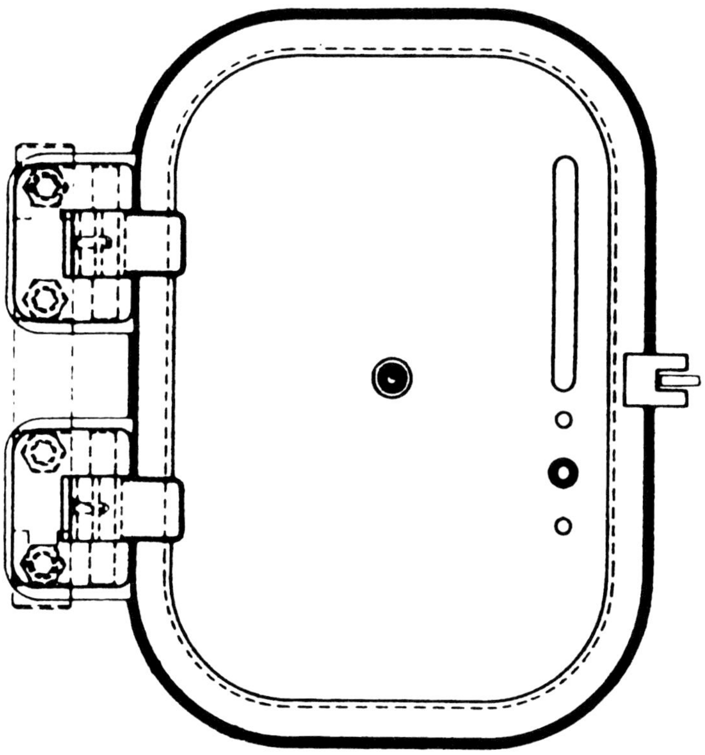

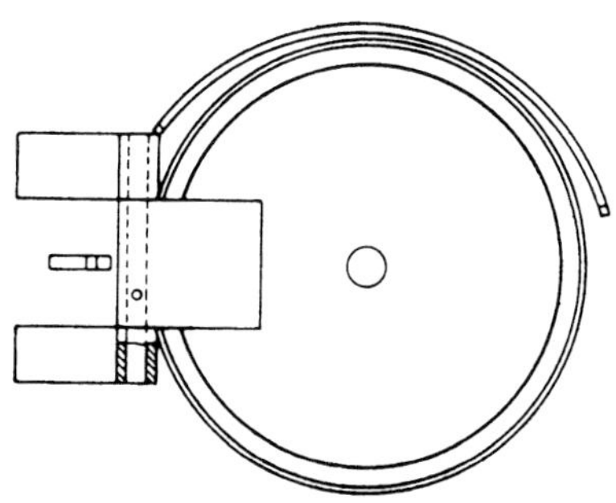

The **Deckel z. Huelsenauswurfloch** (cover over the spent cartridge ejection port) was the same for both the **Serienturm** and **Turm Nr.1-50**. A curved rain guard was welded to the turret roof.

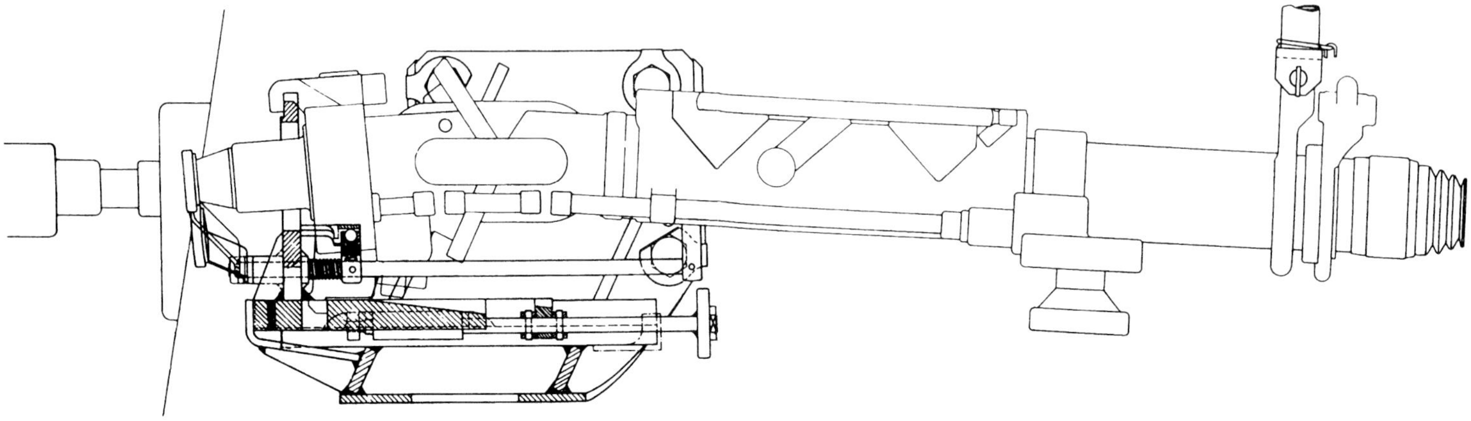

The monocular **T.Z.F.9d** articulated telescopic gun sight mounted in the **Turmzielfernrohrlagerung (021 B 50615)** in the **Serienturm**. A **Wischer** for wiping off the front glass was operated by turning a short rod on the right side. This served more as a cover to protect against dirt and condensation when the sight was not in use than as an effective method of cleaning off the front glass.

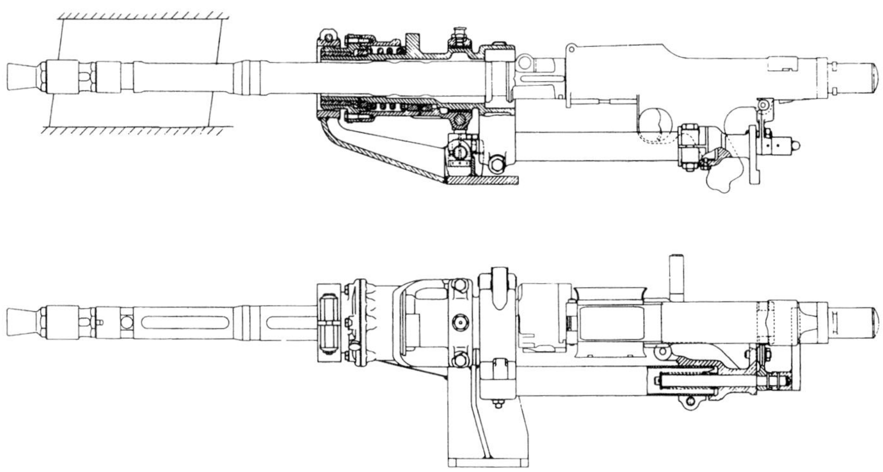

The barrel of the co-axially mounted **M.G.34** protruded through an aperture in the front plate which was shielded by the side of the gun mantlet. The spring buffered mount helped to alleviate stoppages.

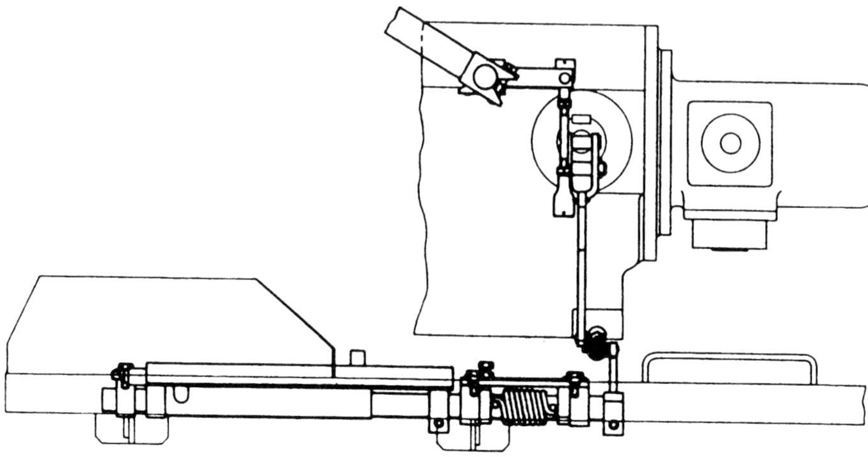

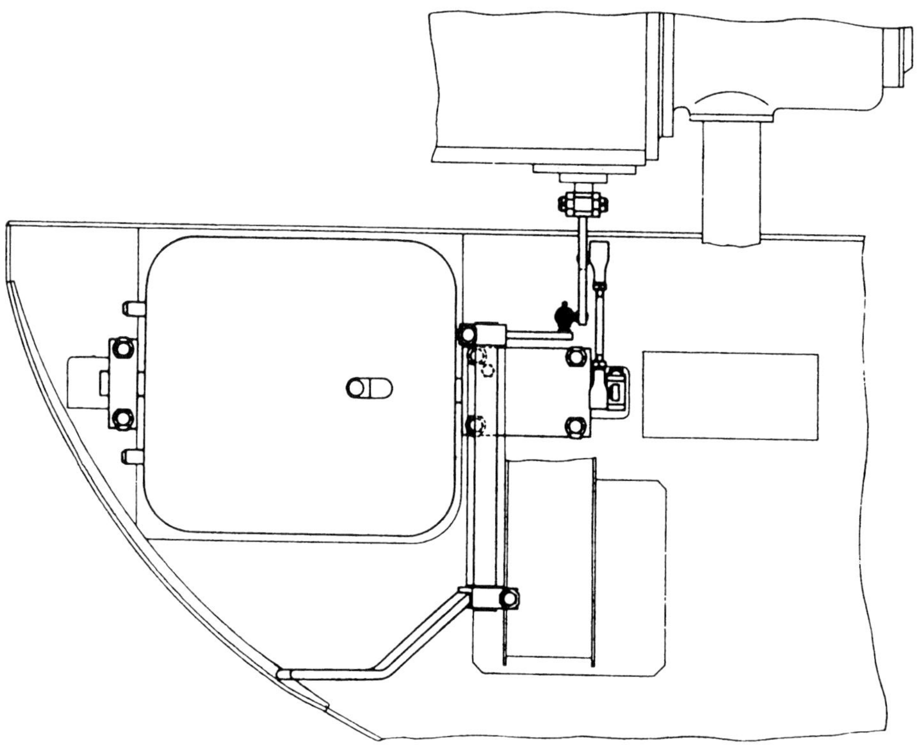

ABOVE: The foot pedal and hand lever controls for the power traverse hydraulic drive. A curved sheet metal guard kept the gunner's feet from protruding beyond the platform.

OPPOSITE: An air compressor was mounted below the gunner's seat and a compressed air tank was mounted in rear-center of the turret platform. An air line connected the compressed air tank to a valve on the breach of the gun. After firing, compressed air automatically blew the remaining burnt propellant fumes out of the gun.

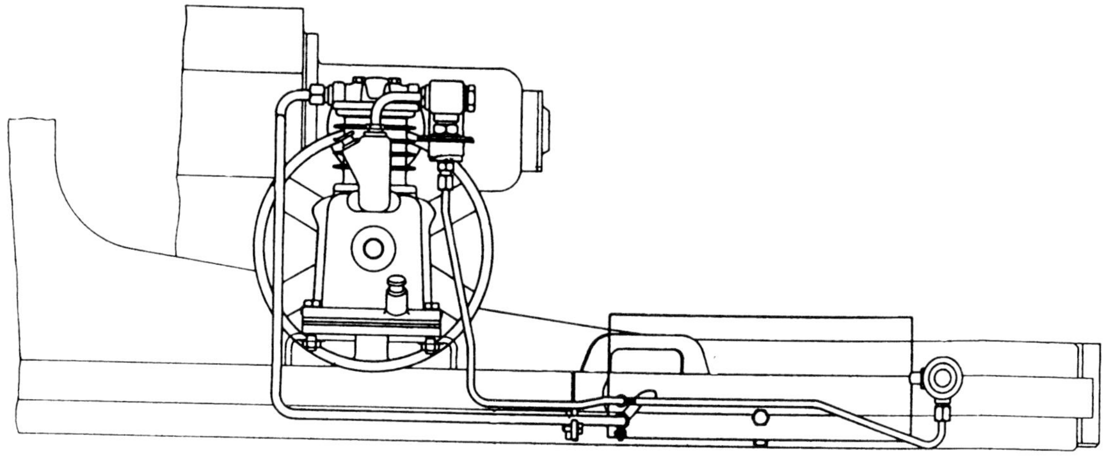

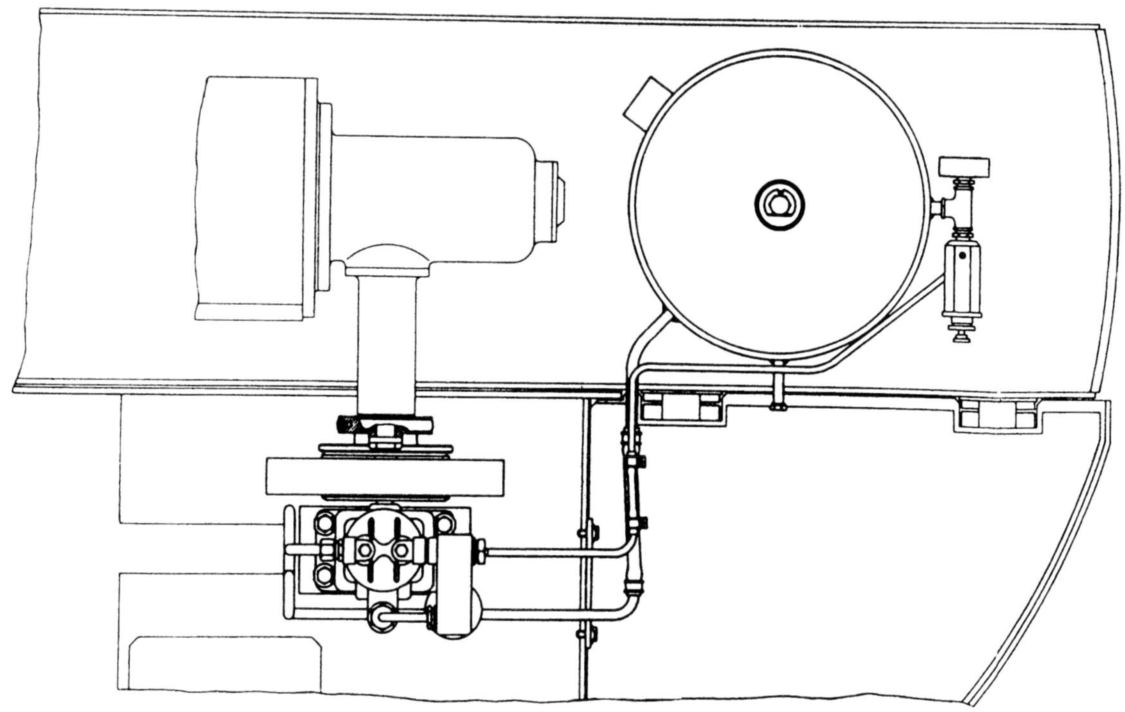

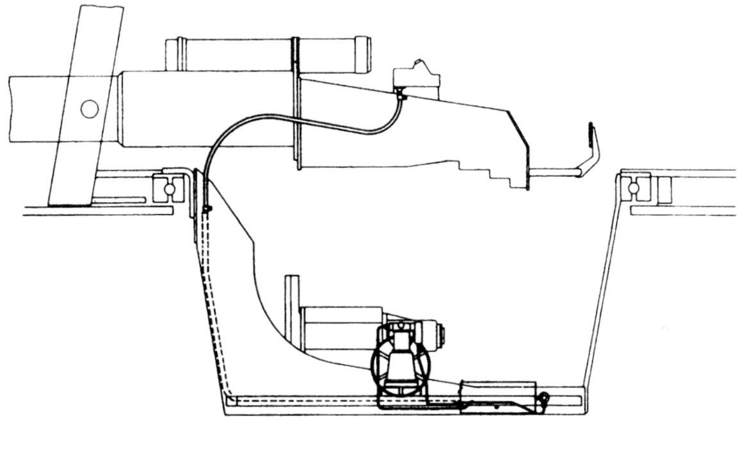

THIS PAGE AND OPPOSITE: Arrangement of components on the turret platform of the **Serienturm**.

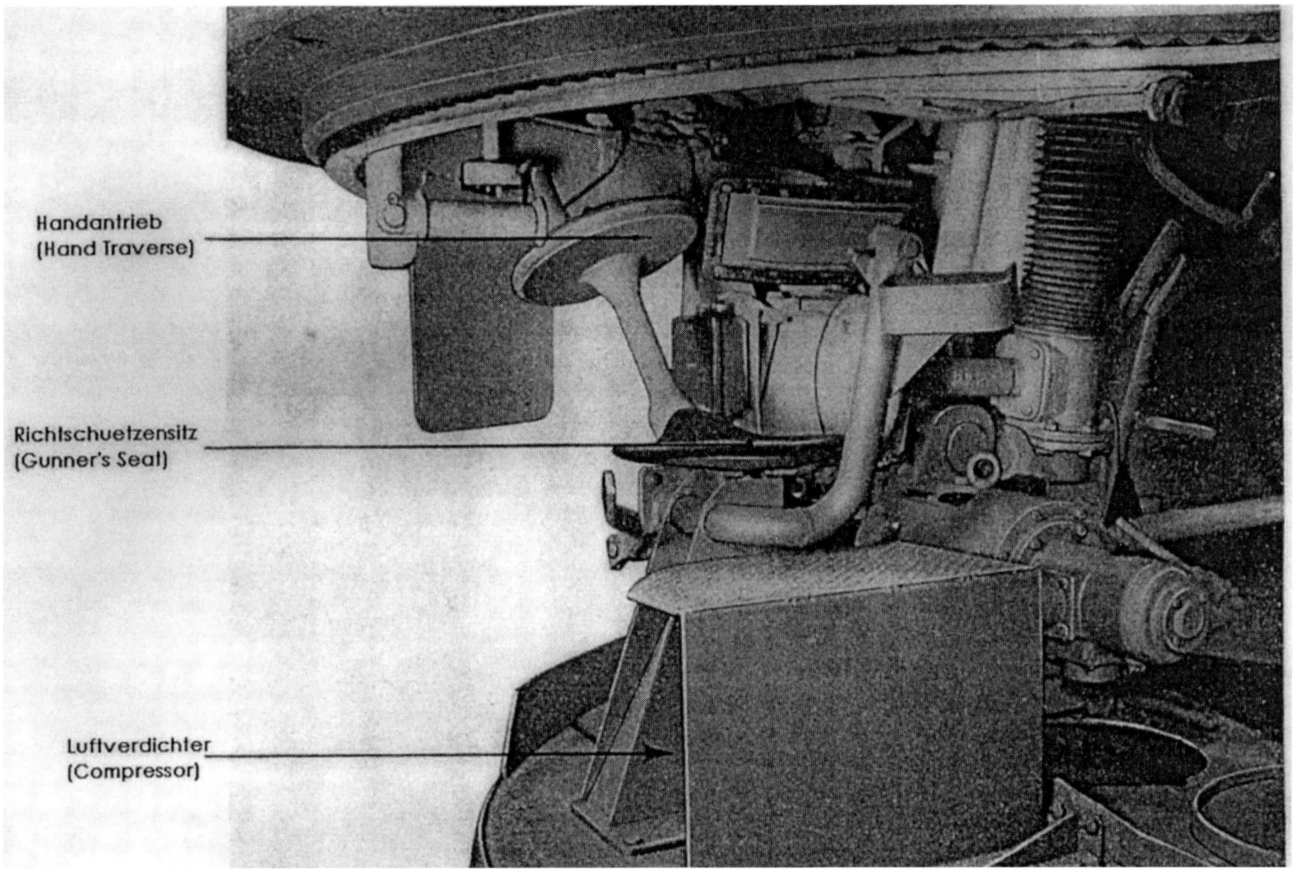
Handantrieb
(Hand Traverse)
Richtschuetzensitz
(Gunner's Seat)
Luftverdichter
(Compressor)

Luftverdichter
(Compressor)

THIS PAGE AND OPPOSITE: Twenty-two rounds for the main gun were stored in the **Geschuetzmunitionslagerung (021 B 50619)** at the rear of the turret. A sheet-metal shield protected the ammunition from fragments spalled off the inside of the turret wall when hit by AP projectiles. Each round was held in place by spring-loaded spacers. Only the inner rounds on the two lower racks were held in place by two straps with quick-release clamps.

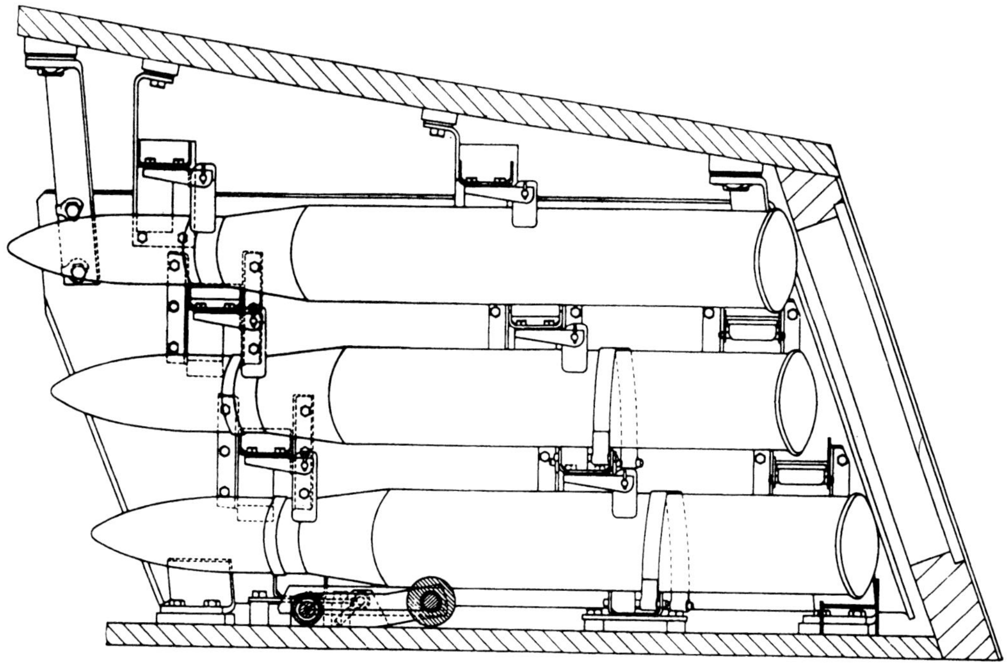

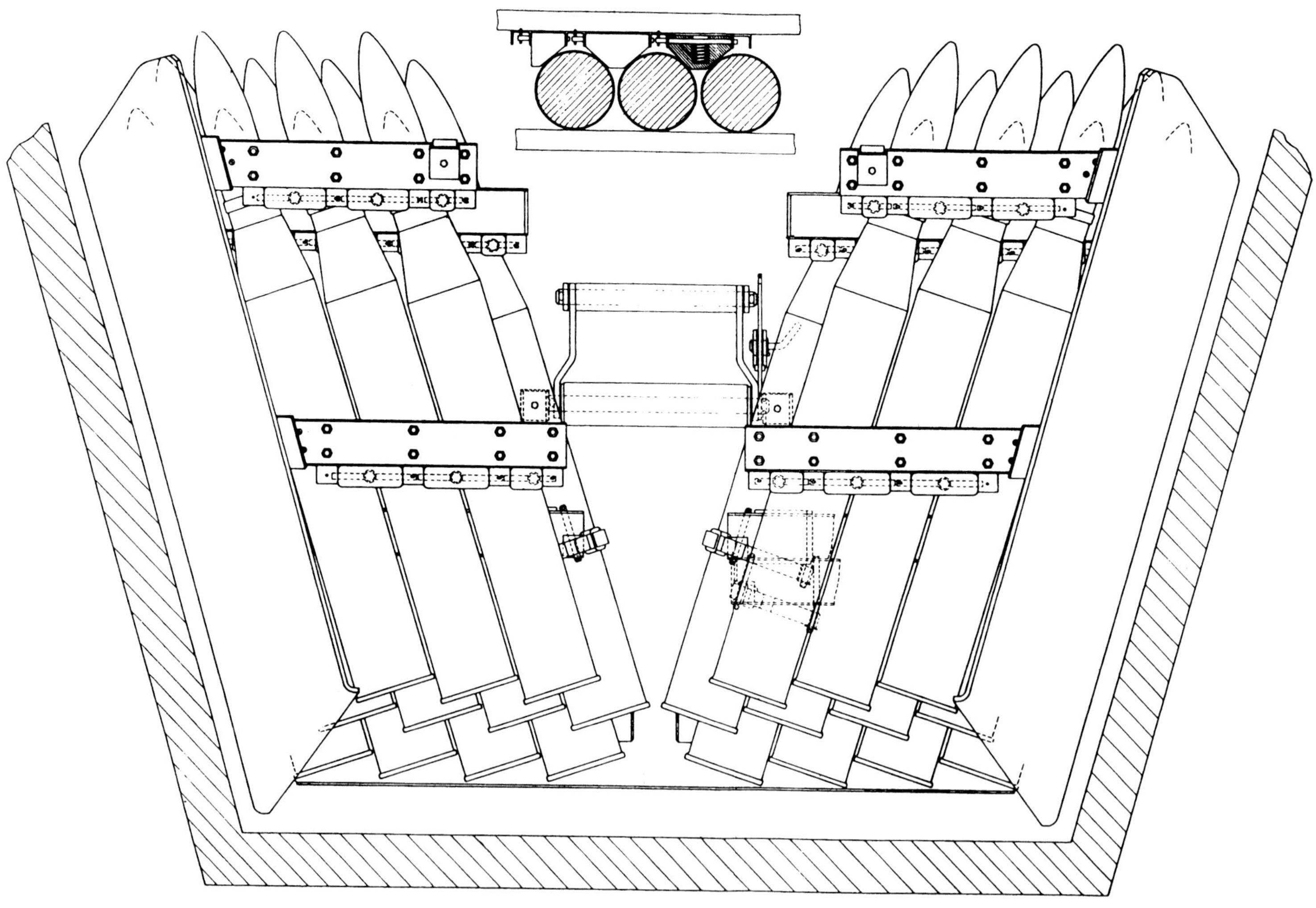

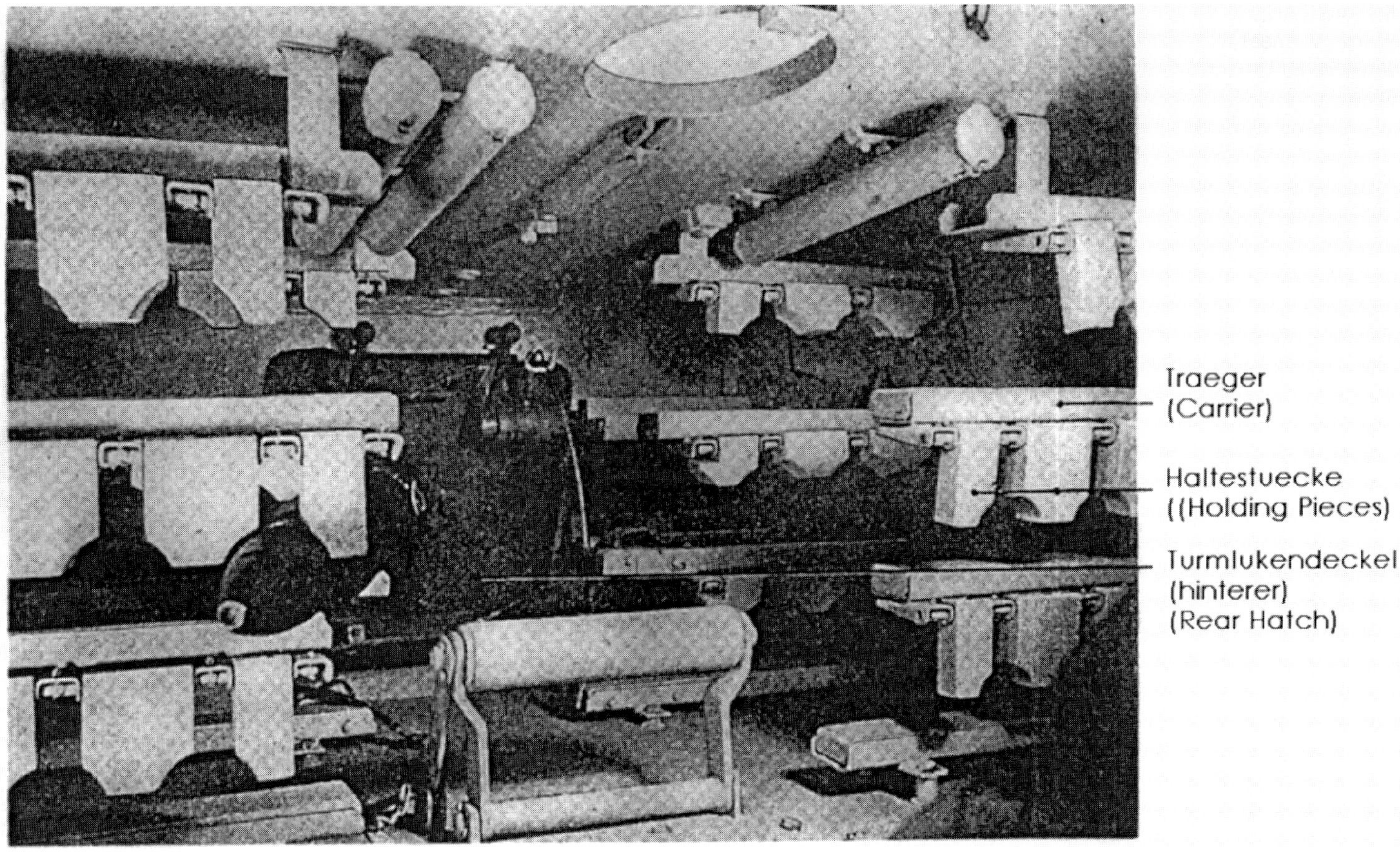
Traeger
(Carrier)
Haltestuecke
((Holding Pieces)
Turmlukendeckel
(hinterer)
(Rear Hatch)

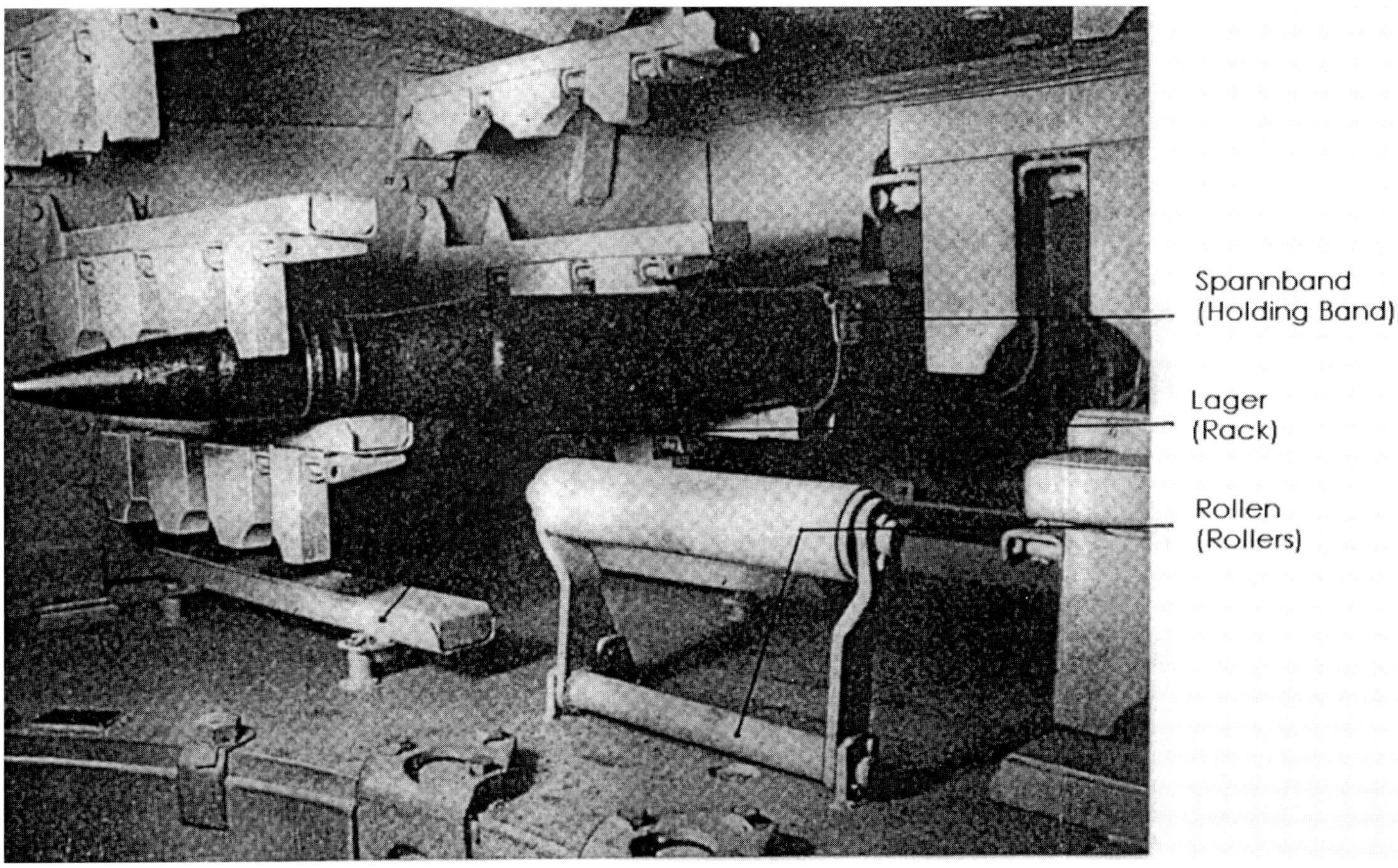
Spannband
(Holding Band)
Lager
(Rack)
Rollen
(Rollers)

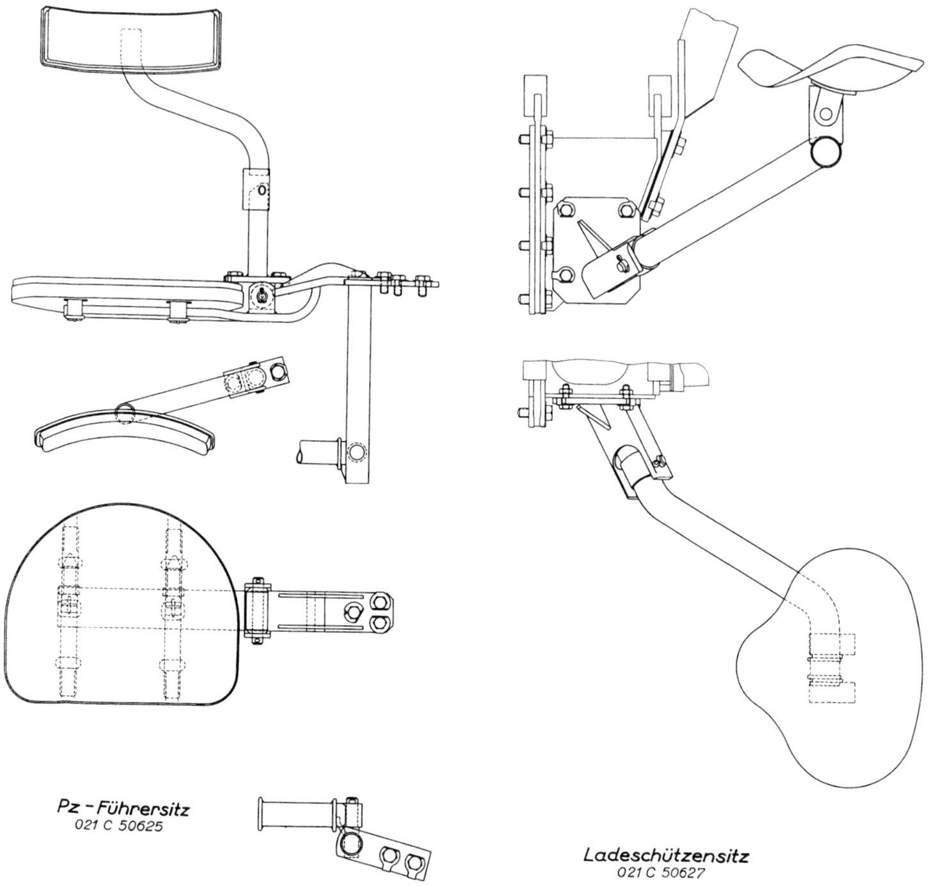

Pz-Fuehrersitz (021 C 50625) commander's seat, **Richtschuetzensitz (021 B 50626)** gunner's seat, and **Ladeschuetzensitz (021 C 50627)** loader's seat in the **Serienturm**. One peg-shaped foot rest for the commander was bolted to the turret ring; the second box-shaped foot rest was welded to the back of the gunner's seat.

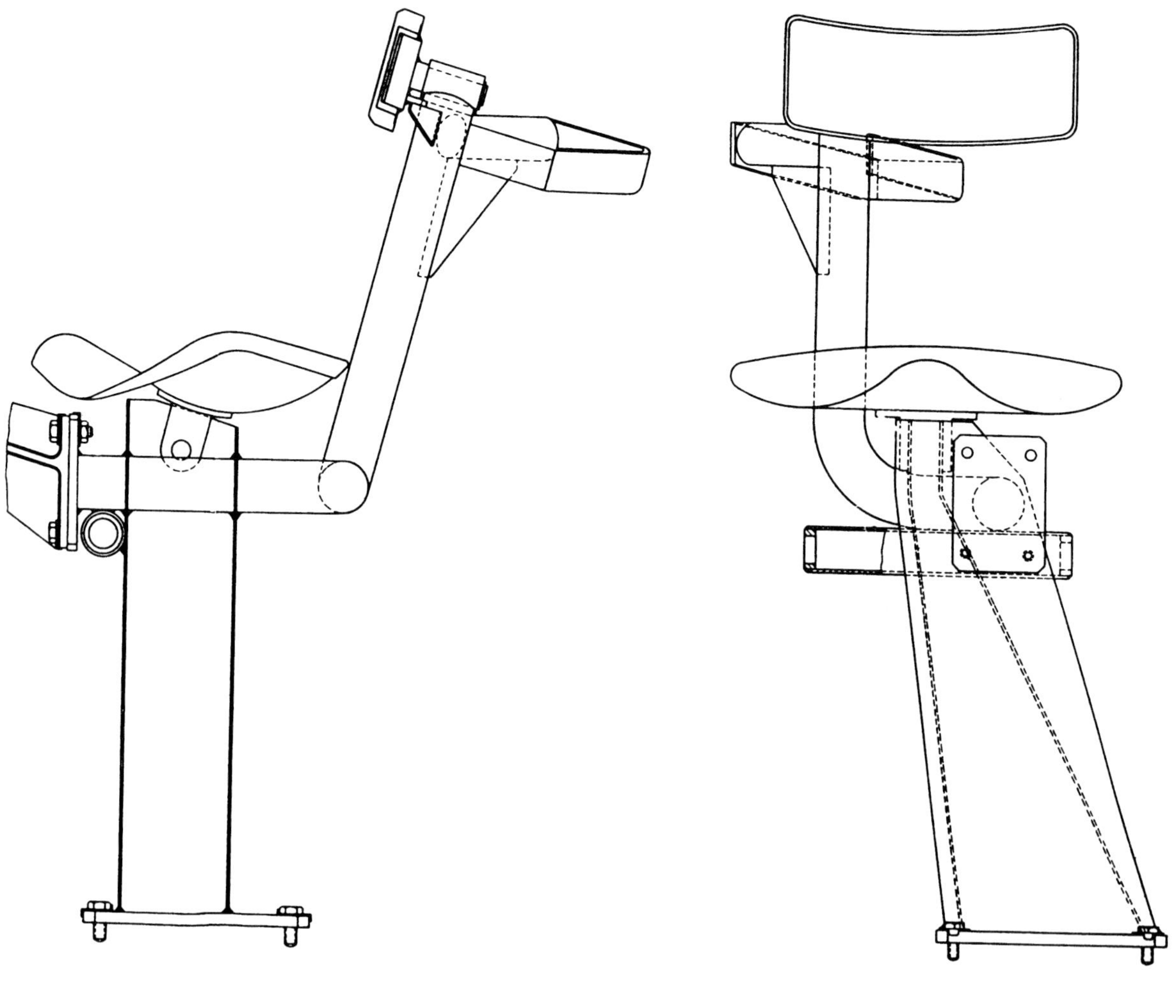

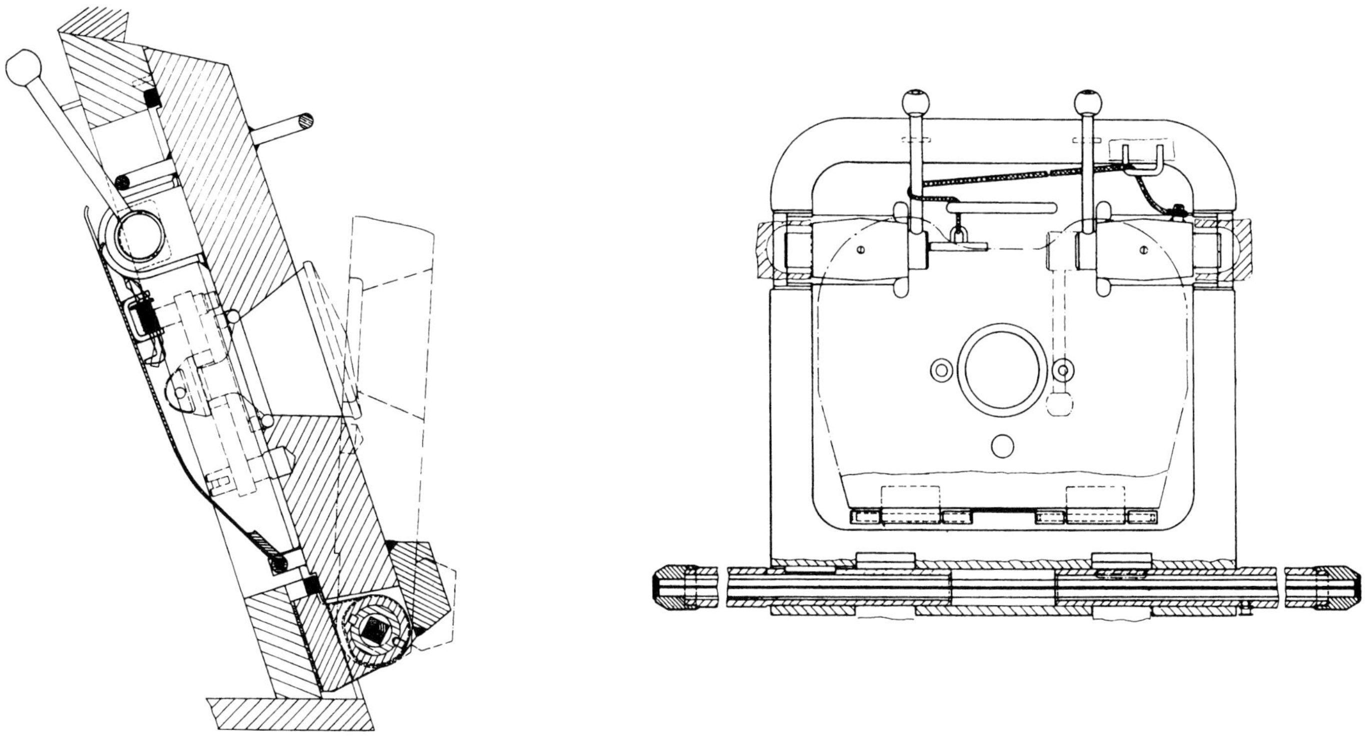

A cross-section view and inside view of the **Turmlukendeckel, hinterer (021 B 50607)** (rear turret hatch) on the **Serienturm** showing details of the torsion bars, locks, sheet metal slide, and pistol port.

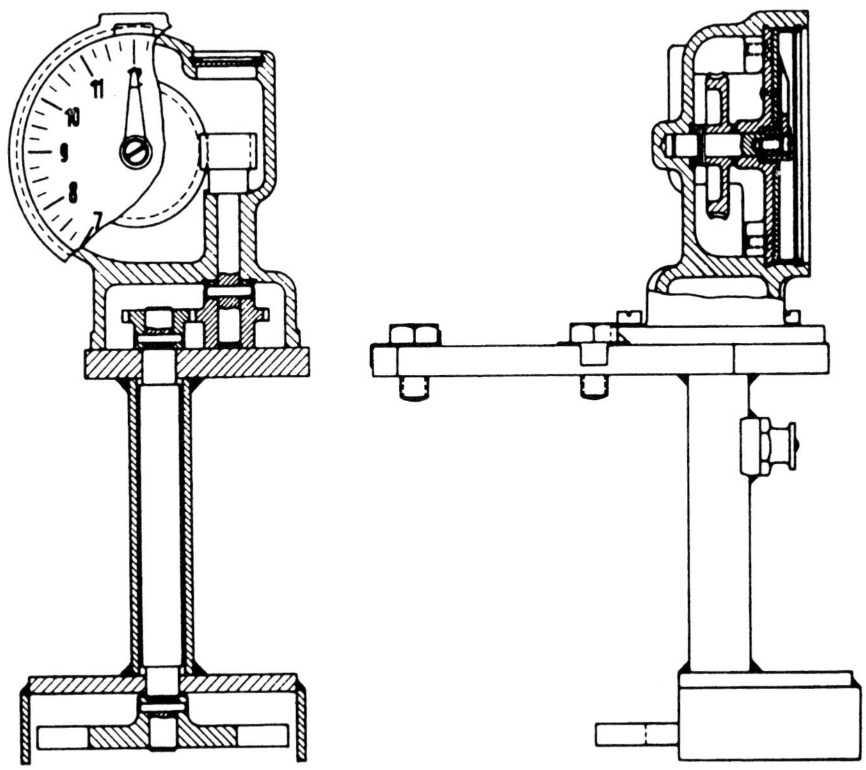

The **Zwoelfuhrzeiger (021 B 50617)** introduced with the **Serienturm** had a double ring dial, marked in both mils and hours. The **Zwoelfuhrzeiger (021 B 50622)** (azimuth indicator), introduced by September 1944, was located on the turret ring so that it could be read by both the gunner and commander.

Panzerbefehlswagen Tiger Ausf.B – completed in June/July 1944 – hangers for spare track links on turret sides – 2-meter rod antenna on turret roof for **Fu 5** radio set – star antenna on rear deck on porcelain insulator inside armor pot for **Fu 8** radio set – cylindrical spare antenna rod stowage container across top of hull rear.

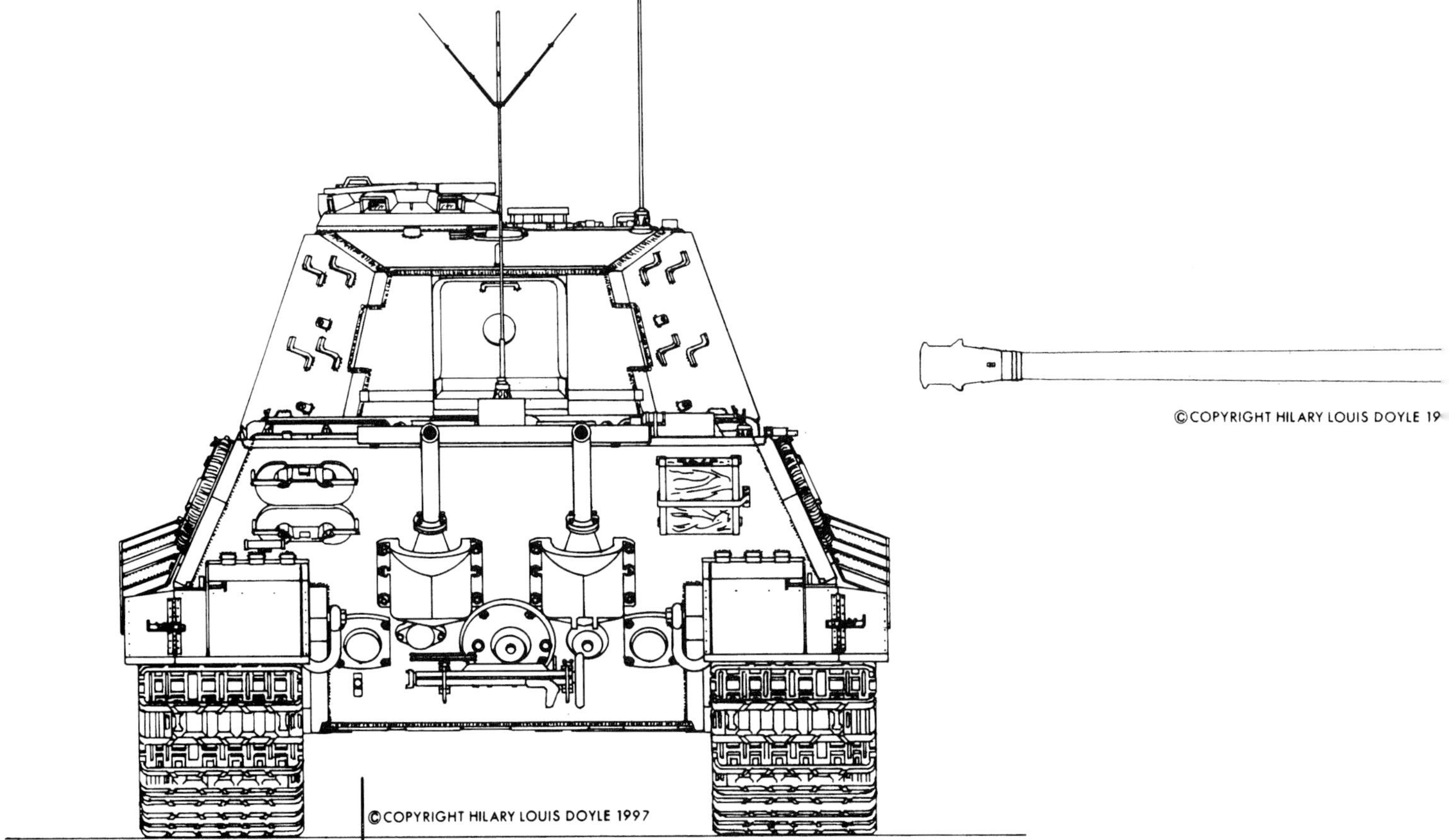

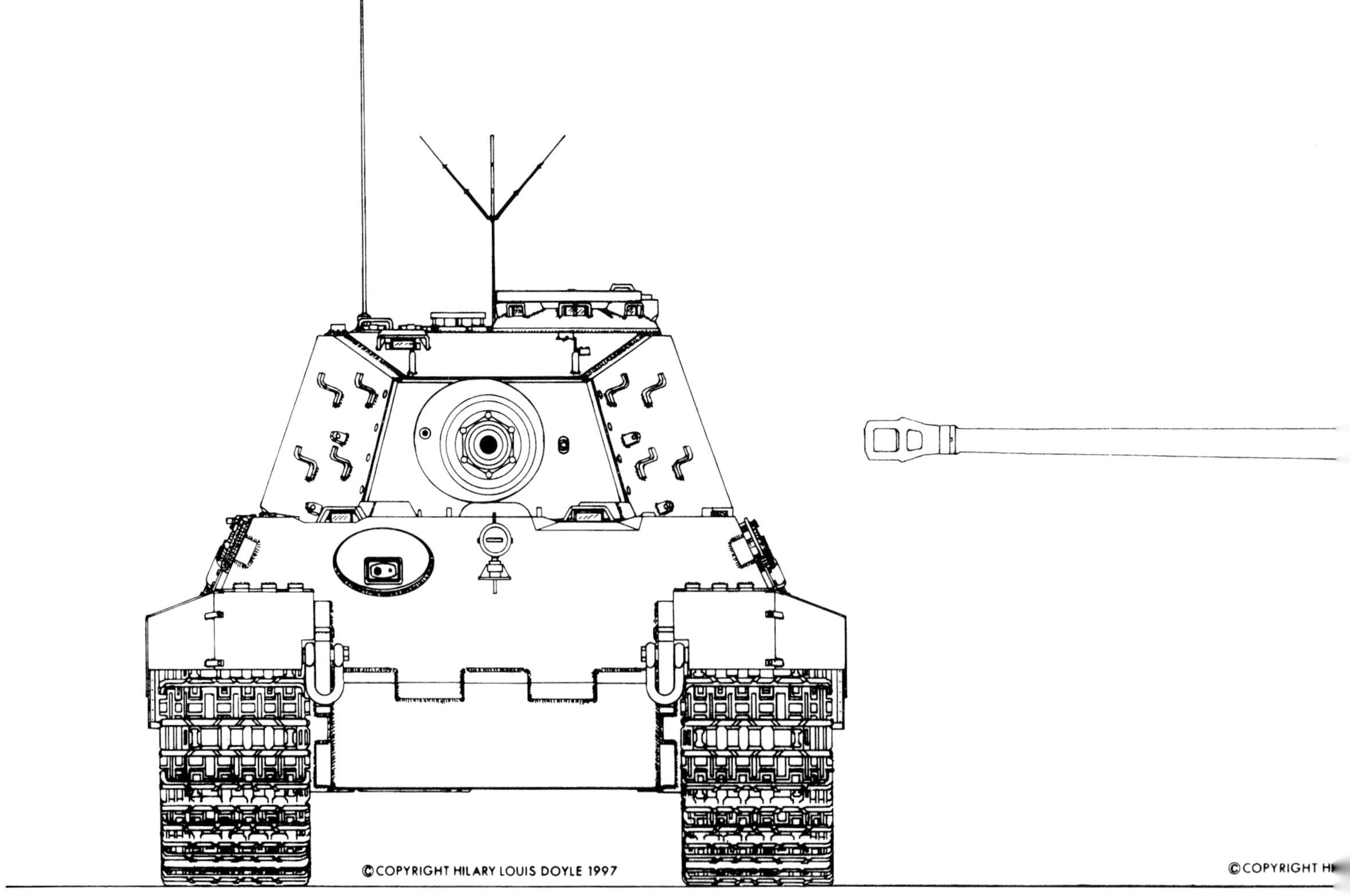

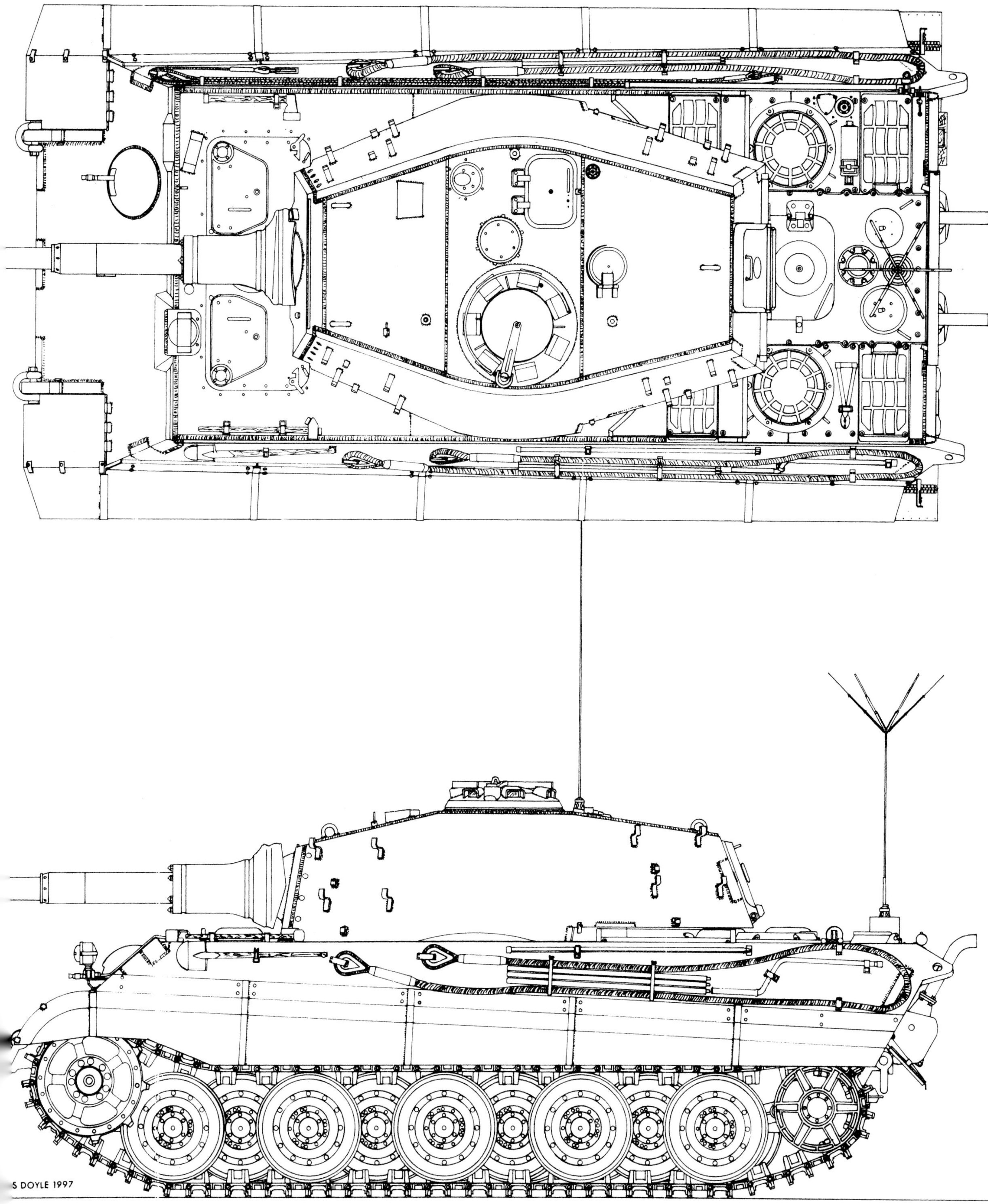
S DOYLE 1997

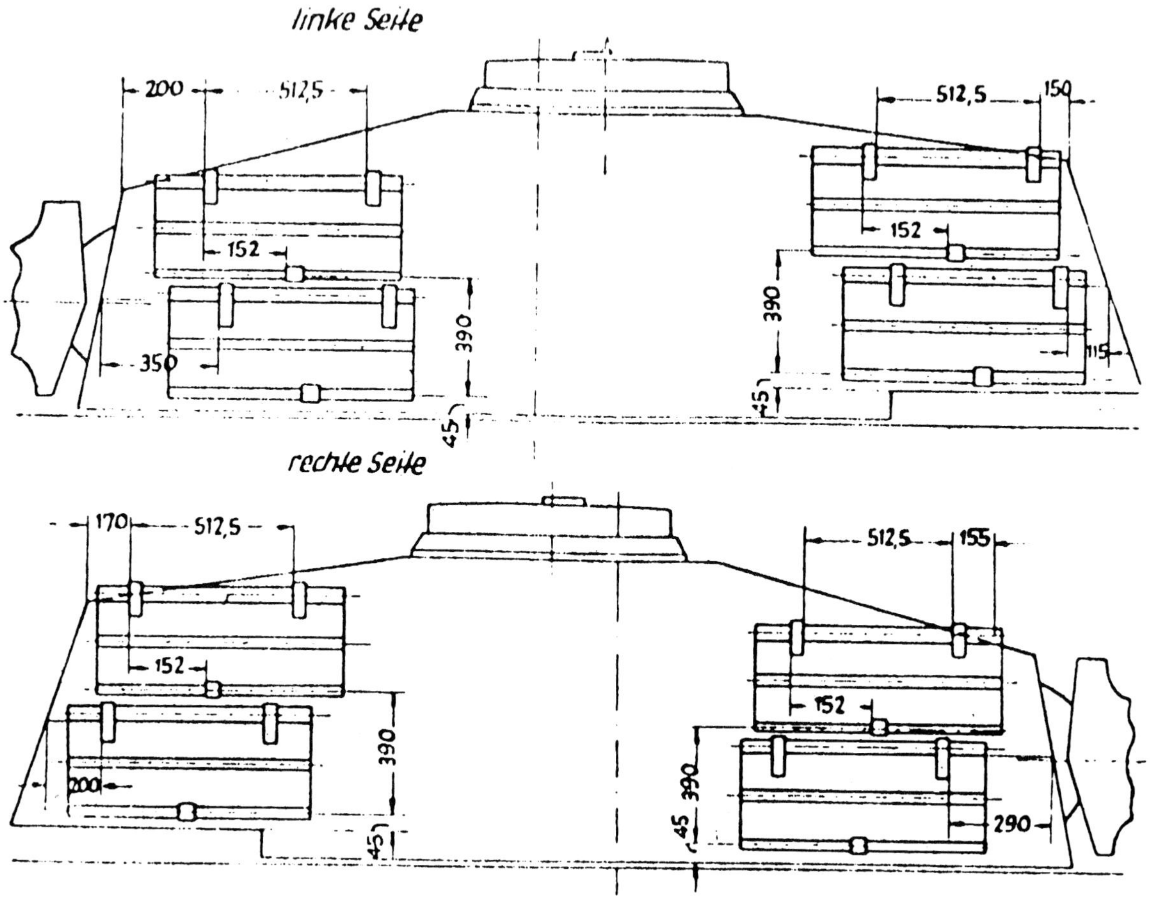

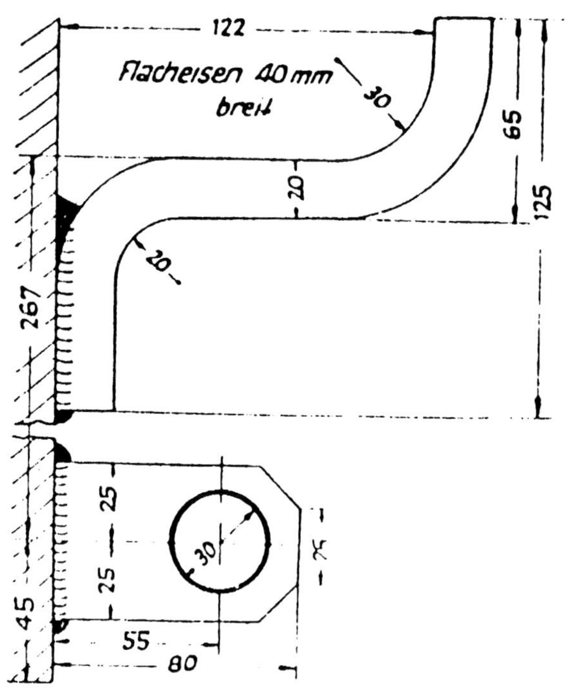

Starting in early July 1944, spare track link hangers were welded to the sides of the **Serienturm** as shown on the drawings.

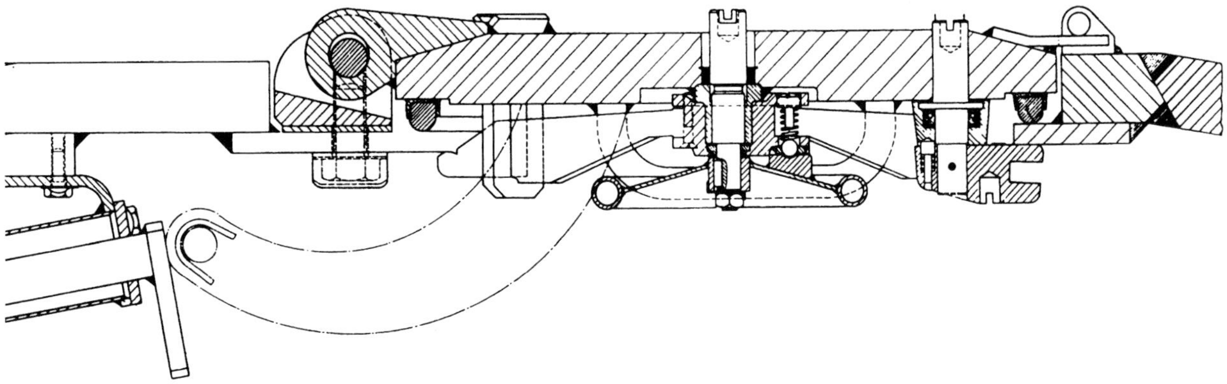

Turmlukendeckel oberer
Spae 104-22
nach 021B 50640

A 40 mm thick **Turmlukendeckel, oberer (021 B 50640)** was introduced on the **Serienturm** starting in early July 1944. The purpose of the four U-shaped channels, welded onto each corner, is not described in the turret manual. Only remnants of the welds remain where they were cut off prior to final assembly of the **Serienturm**.

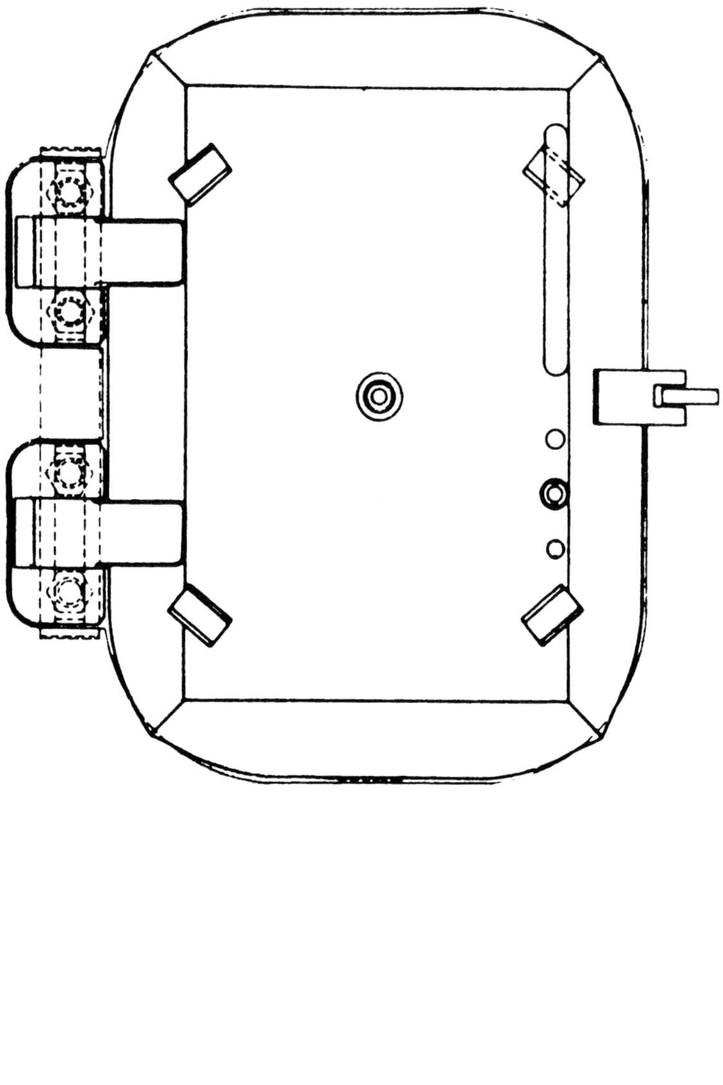

Details of the drain covers and escape hatch on the bottom of the hull are seen in this photo of a **Tiger Ausf.B** being paid a visit by General Eisenhower. The trailing roadwheel suspension arms on the right side and leading roadwheel suspension arms on the left side are also revealed. (NA)

This **Tiger Ausf.B** with **Serien-Turm** was completed at Henschel in late June 1944 and issued to the **3.Kompanie/schwere Panzer-Abteilung 503**. It had the new 40 mm thick loader's hatch, three **Pilze** welded onto the turret roof for the **Behelfskran 2t**, and the lighter muzzle brake for the **8.8 cm Kw.K.43 L/71**, but didn't have spare track hangers welded onto the turret sides. (BA)

These factory shots of a **Tiger Ausf.B** with 660 mm wide **Verladekette** (transport track) were used as illustrations in a manual dated 1 July 1944. Completed by Henschel in June 1944, it has the three **Pilze** welded onto the turret roof and spare track hangers welded onto the turret sides, but is missing the armor guards for the torsion bars counterbalancing the rear turret hatch. The positions of the roadwheels on each side (due to the spacing of the torsion bars, the left side was ahead of the right) can easily be recognized in these photos. (NA)

A very rare shot of the **Behelfskran 2t**, mounted in the three **Pilze** on the turret roof of this **Tiger Ausf.B**, being used to remove the rear deck over the engine compartment. The loader's hatch is still the 15 mm thick forging. (BA)

Shown after capture, **Tiger Ausf.B** (**Fgst.Nr.280093**, **Turm Nr.280124**), completed by Henschel on or about 18 July 1944, was issued to the **1.Kompanie/schwere-SS-Panzer-Abteilung 101**. Now on display at the College of Military Science in Shrivenham, England, it has the 40 mm thick loader's hatch and the welded base commander's cupola along with three **Pilze** welded to the turret roof. But it didn't have mounts for the poison gas recognition panels welded onto the top of the gun mantlet and rear of the turret roof. (TTM)

Closeup of the right front corner of **Tiger Ausf.B** (**Fgst.Nr.280100**), completed by Henschel on or about 23 July 1944, showing details of the mudguards and the stowage of the 15 mm thick, 15 meter steel cable used for replacing the tracks. (WJS)

Tiger Ausf.B (**Fgst.Nr.280101**, **Turm Nr.280110**) when it was displayed at the Ordnance Museum in Aberdeen, Maryland prior to returning it to Germany where it is now on display at the Panzer Museum in Munster. It had been issued to the **1.Kompanie/schwere-SS-Panzer-Abteilung 101**. When captured, it was found blown onto its back with the gun tube broken and out of battery. (Disregard the unit tactical sign). (APG)

THIS PAGE AND OPPOSITE: Details of **Tiger Ausf.B** (**Fgst.Nr.280101**, **Turm Nr.280110**), now on display at the Panzer Museum in Munster, Germany. (Disregard the camouflage pattern.) (TLJ)

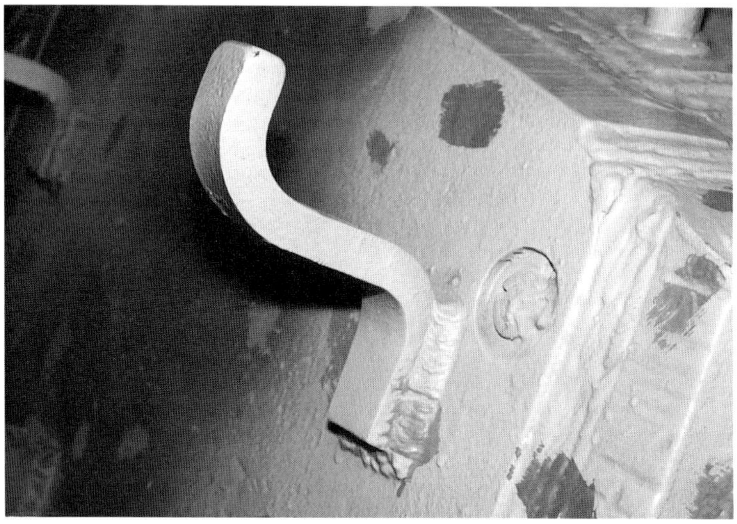

Track hangers on the right turret side.

Commander's cupola welded to the turret roof. The stanchions for the **Regenschutz** (canvas rain guard) are located directly in front of guard for the pivoting arm and directly behind the right periscope guard. (A bolt has been inserted into the right stanchion.)

The double wire welded to the front periscope guard, used by the commander to guide the gunner onto target.

Looking down the barrel of the **Nahverteidigungswaffe** at the recess for the cartridge and hole for the firing pin.

The stepped aperture cut for the monocular gun sight into the turret front plate.

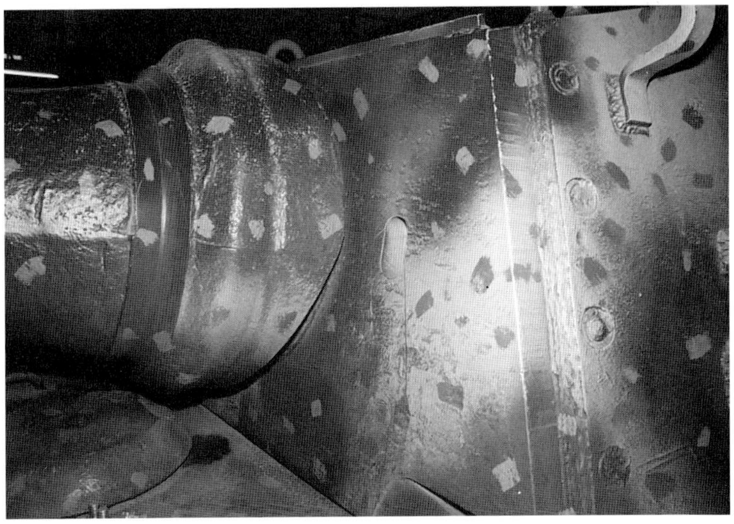

The rear upper edge of the armor guard covering the hull vent was cut away to allow clearance for traversing a fully depressed gun.

The stepped aperture in the armor guard for the hull machinegun ball-mount that was welded to the glacis plate. The smaller hole was for the **K.Z.F.2** sight.

The weight of the rear hatch lid was suspended by two torsion bars. Armor guards for the torsion bars weren't introduced on the **Serienturm** until August 1944. A recess was cut into the turret rear wall for fitting the rear hatch cover.

The port on the left was for track adjustment. Underneath this port are the remains of a reflector. The port under the exhaust pipe guard was for the **Kuhlwasserheizgeraet** (engine coolant heater). Only remnants remain of the jack hangers that were welded to the lower hull rear.

Two palls and the offset port (in the large cover to the rear of the engine) were for the **Kurbelwellen-Benzinanlasser** (gasoline engine starter). A crank for turning the engine by hand was fed through a ring pivoting on the two eyelets on the exhaust pipe armor guard and inserted into the small port just below the right-hand exhaust pipe.

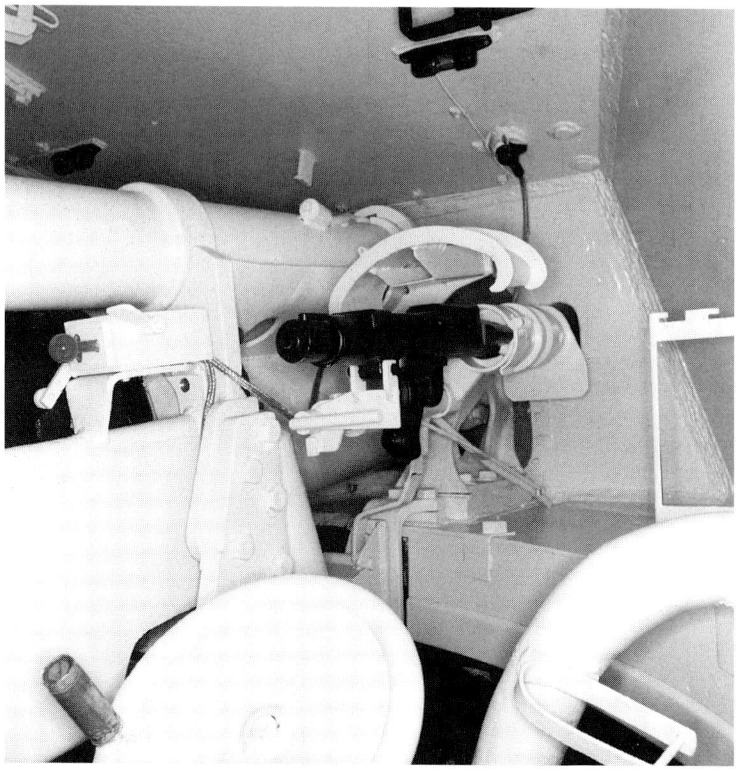

LEFT: The left front corner of the renovated turret of **Tiger Ausf.B** (**Fgst.Nr.280101**, **Turm Nr.280110**). The dial of the azimuth indicator to the gunner's left front was driven by gears directly from the turret ring gear. The rear of the monocular **T.Z.F.9d** telescopic gun sight was supported from an adjustable hanger suspended from the turret roof. (WJS)

RIGHT: The right front corner of the renovated turret of **Tiger Ausf.B** (**Fgst.Nr.280101**, **Turm Nr.280110**). Ammunition was fed into the left side of the **M.G.34** by suspending the bag on the right and feeding the belt over the curved tray. (WJS)

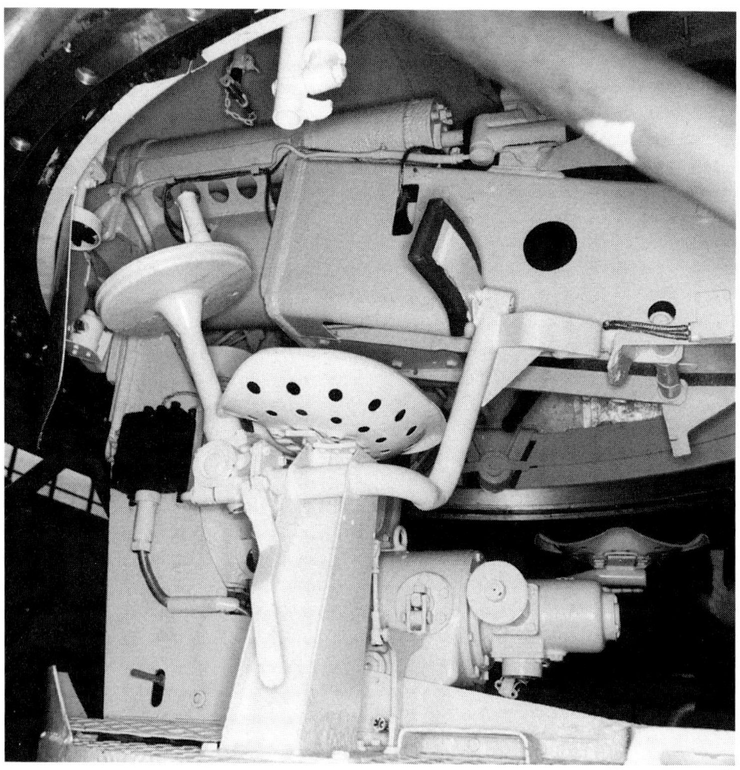

LEFT: The gunner's position in the renovated turret of **Tiger Ausf.B** (**Fgst.Nr.280101**, **Turm Nr.280110**). The sheet metal guard hung from the turret ring and the guard welded to the turret platform protected the gunner's knees and toes. The gunner's feet rested on a pivoting plate for controlling the hydraulic traverse. Tilting the plate down on the right traversed the turret to the right, and left, left. The crooked lever beside the gunner's seat support was also used for controlling the turret traverse with the gunner's left hand. (It isn't adjusted to the correct position.) (WJS)

RIGHT: Looking forward through the open rear hatch into the renovated turret of **Tiger Ausf.B** (**Fgst.Nr.280101**, **Turm Nr.280110**). The wooden spacers in the empty ammunition racks are hanging down instead of being held out of the way with springs as they were originally designed. The rollers on the deck were installed as an aid to loading the gun. A cable, used for pulling the rear hatch closed, was fed through the loop welded to the opening. (WJS)

The roof of **Turm Nr.280066** on **Tiger Ausf.B** (**Fgst.Nr.280105**), completed by Henschel on or about 27 July 1944. This turret was outfitted with the 15 mm thick loader's hatch, three **Pilze** for the **Behelfskran 2t**, and the commander's cupola with welded base. The cupola lid had already been pilfered, or was it removed by the crew? (NA)

The rear of **Turm Nr.280066** on **Tiger Ausf.B** (**Fgst.Nr.280105**) showing the position of the third **Pilz**, the sheet metal shield inside the rear hatch, and the lack of armor guards for the torsion bars counterbalancing the turret rear hatch. (NA)

The rear deck of **Tiger Ausf.B** (**Fgst.Nr.280105**) showing the two different shapes for the carburetor air intake vent covers (the forward cover was flattened for turret clearance), the shape of the wire mesh screens over the forward cooling air intake grating, and the screen missing from the circular flange over the engine compartment vent. (NA)

A closeup of the turret front of a **Tiger Ausf.B**, completed by Henschel in late August/early September 1944 and issued to the **schwere Panzer-Abteilung 503**. Apparently, the commander didn't like to fight buttoned up, as the hatch lid for the commander's cupola has been removed from the pivoting arm. (BA)

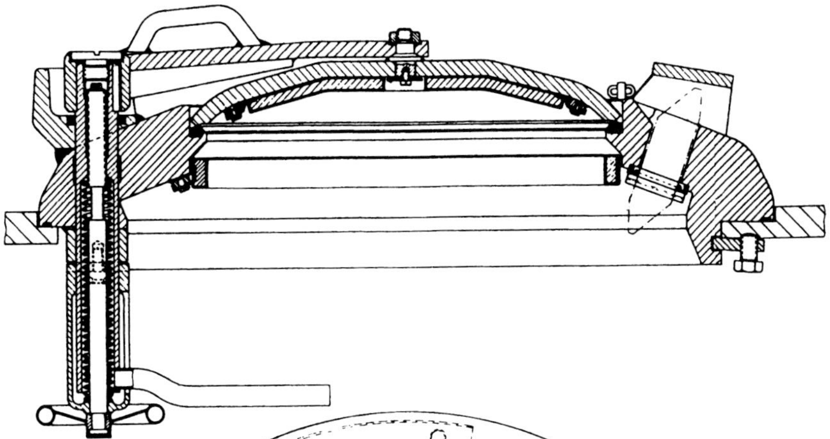

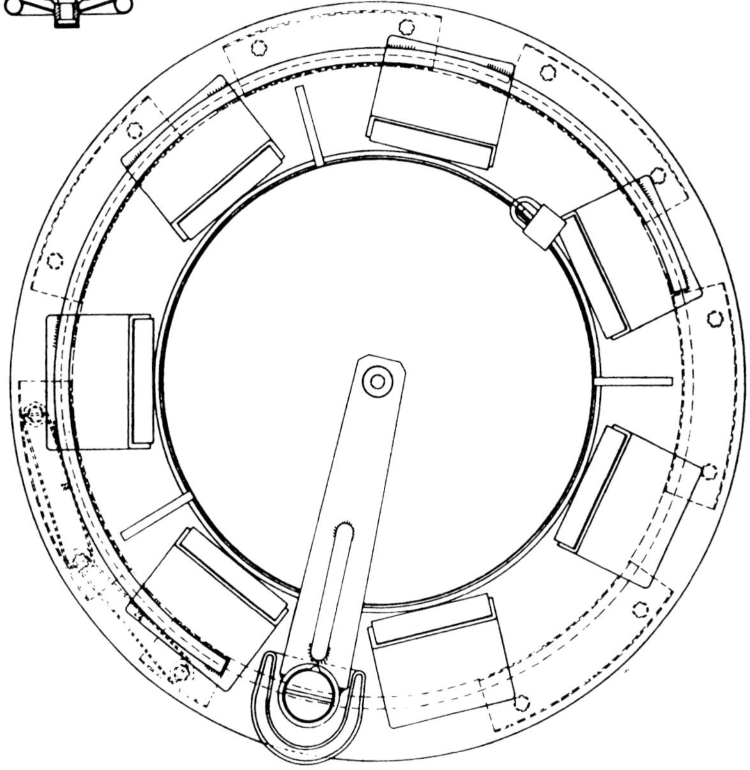

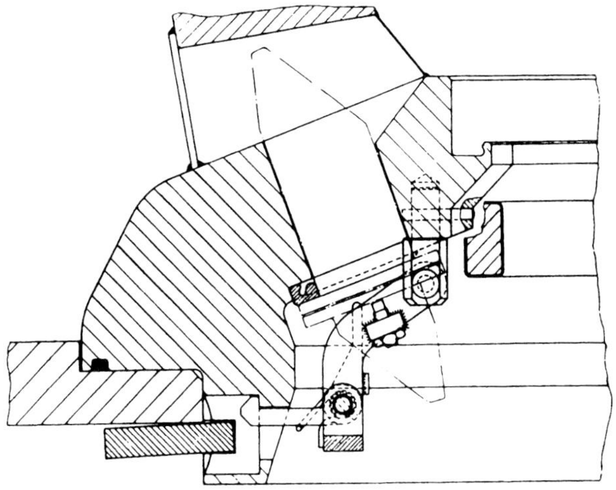

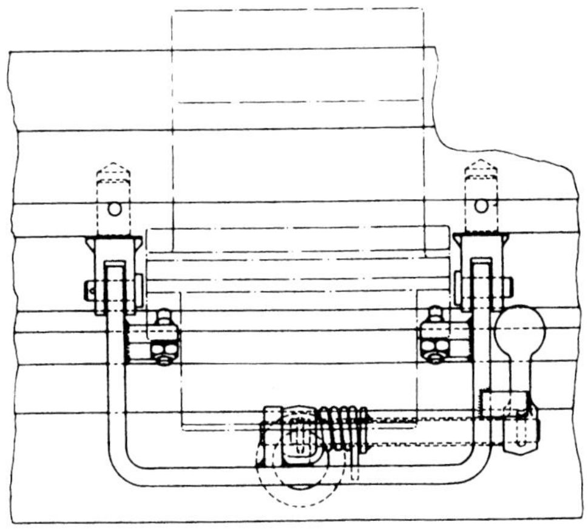

By July 1944, quick-release mounting brackets for the two-piece periscopes were being installed in the **Pz-Fuehrerkuppel (021 B 2762)** commander's cupola. The periscopes could be easily replaced even when the crew were wearing heavy gloves. Starting in August 1944, the **Pz-Fuehrerkuppel (021 B 50641)**, bolted to the turret roof, replaced the welded model. The space was increased inside the guard surrounding the base of the hatch cover support arm.

Details of the rear of a **Tiger Ausf.B**, completed by Henschel in late August/early September 1944 and issued to the **schwere Panzer-Abteilung 503**. Very few changes were made to the tool and equipment stowage layout during the production run of the **Tiger Ausf.B**. The 20-liter "Jerry Cans" hung on the turret rear and the track link bracing the exhaust pipes are both examples of crew mods. (BA)

The turret roof of **Tiger Ausf.B** (**Fgst.Nr.280215**, completed by Henschel on 2 September 1944) with the plug in the **Nahverteidigungswaffe**, the 40 mm thick loader's hatch lid, and the commander's cupola bolted to the turret roof. One of the two stanchions, used for erecting the **Regenschutz** (canvas rain guard) over the commander's cupola, was welded between the periscope guards at 5 o'clock. (HLD)

Tiger Ausf.B (**Fgst.Nr.280215**, completed by Henschel on 2 September 1944) is now on display at the armor school in Thun, Switzerland. Most of the screens and their frames over the gratings, the equipment stowage brackets, and the armor cover for the fuel filler cap are missing. (HLD)

Panzerkampfwagen Tiger Ausf.B – Fgst.Nr.280243 completed in September 1944 – no **Zimmerit** – 40 mm thick loader's hatch – mounts for poison gas identification panels – commander's cupola bolted to turret roof – armor guards for torsion bars on turret rear – armor cover over engine compartment vent on rear deck.

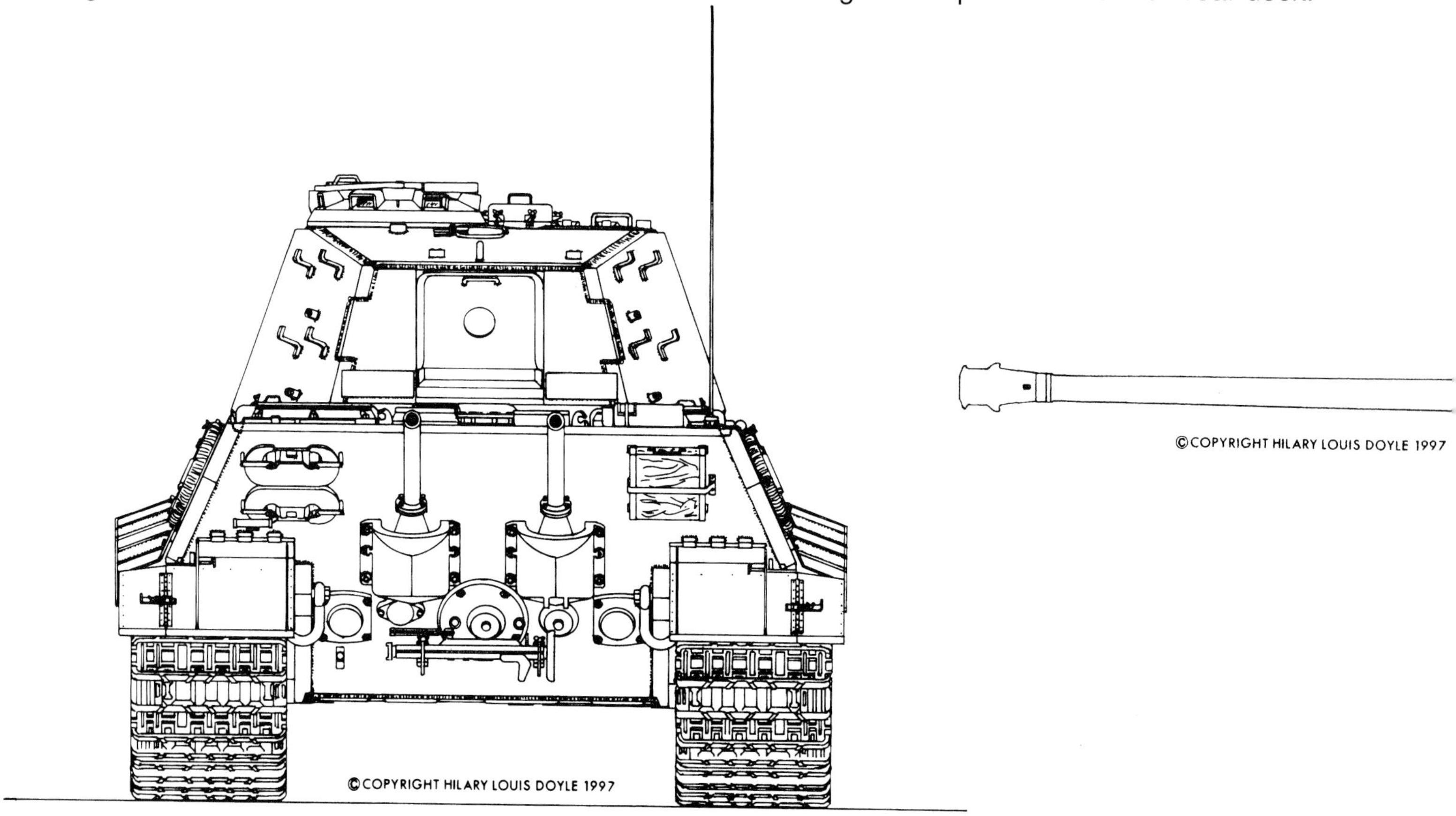

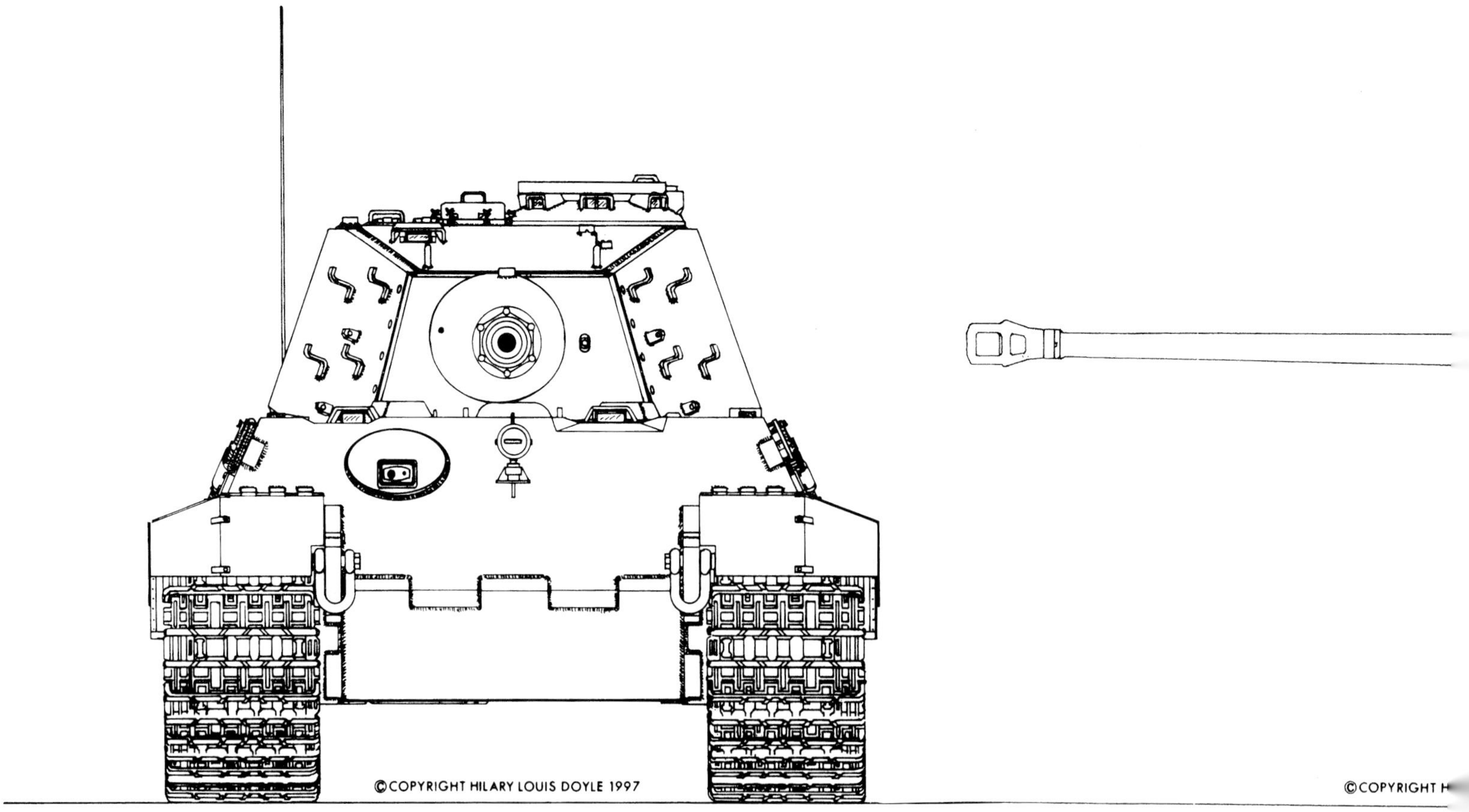

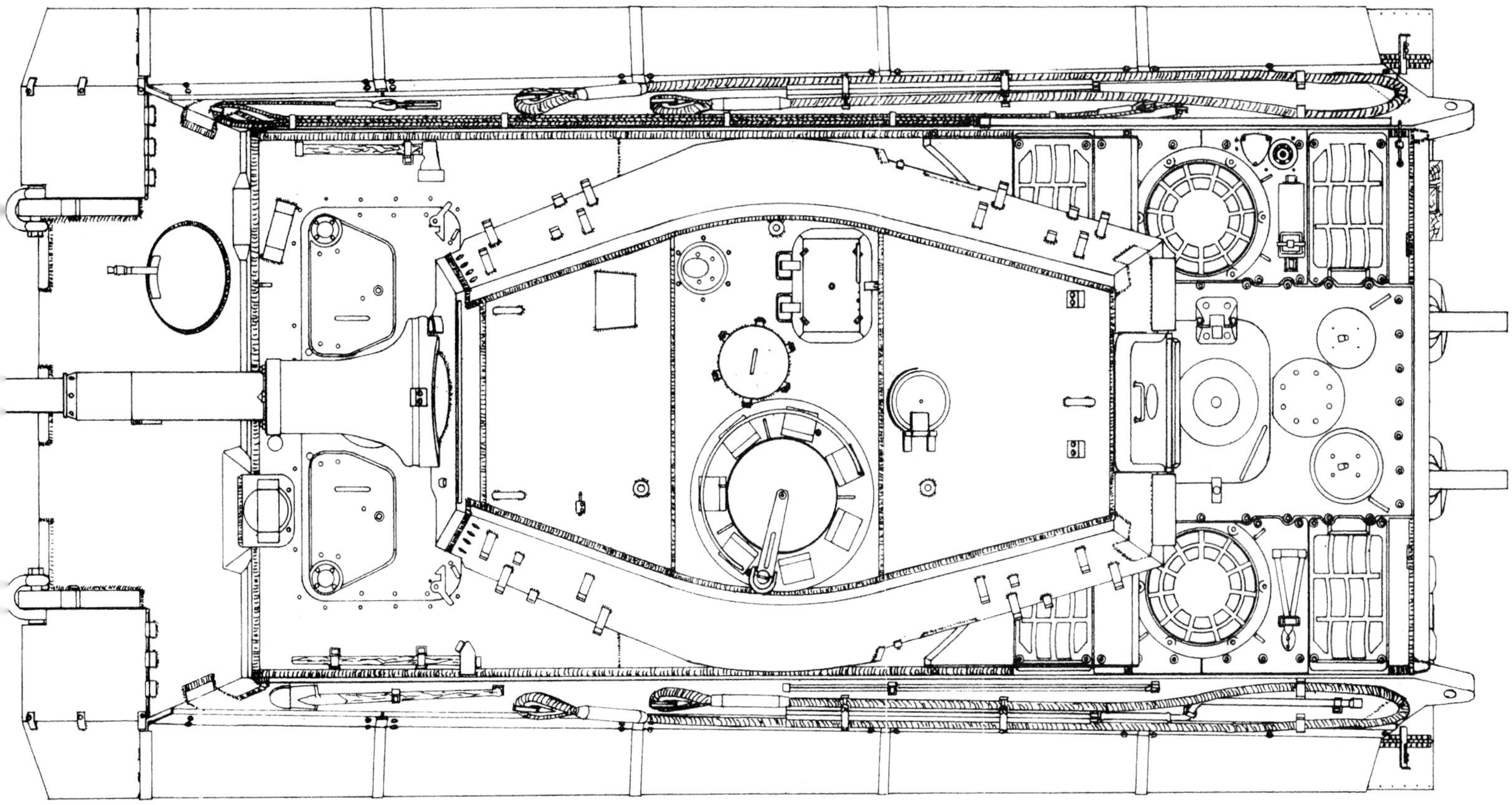

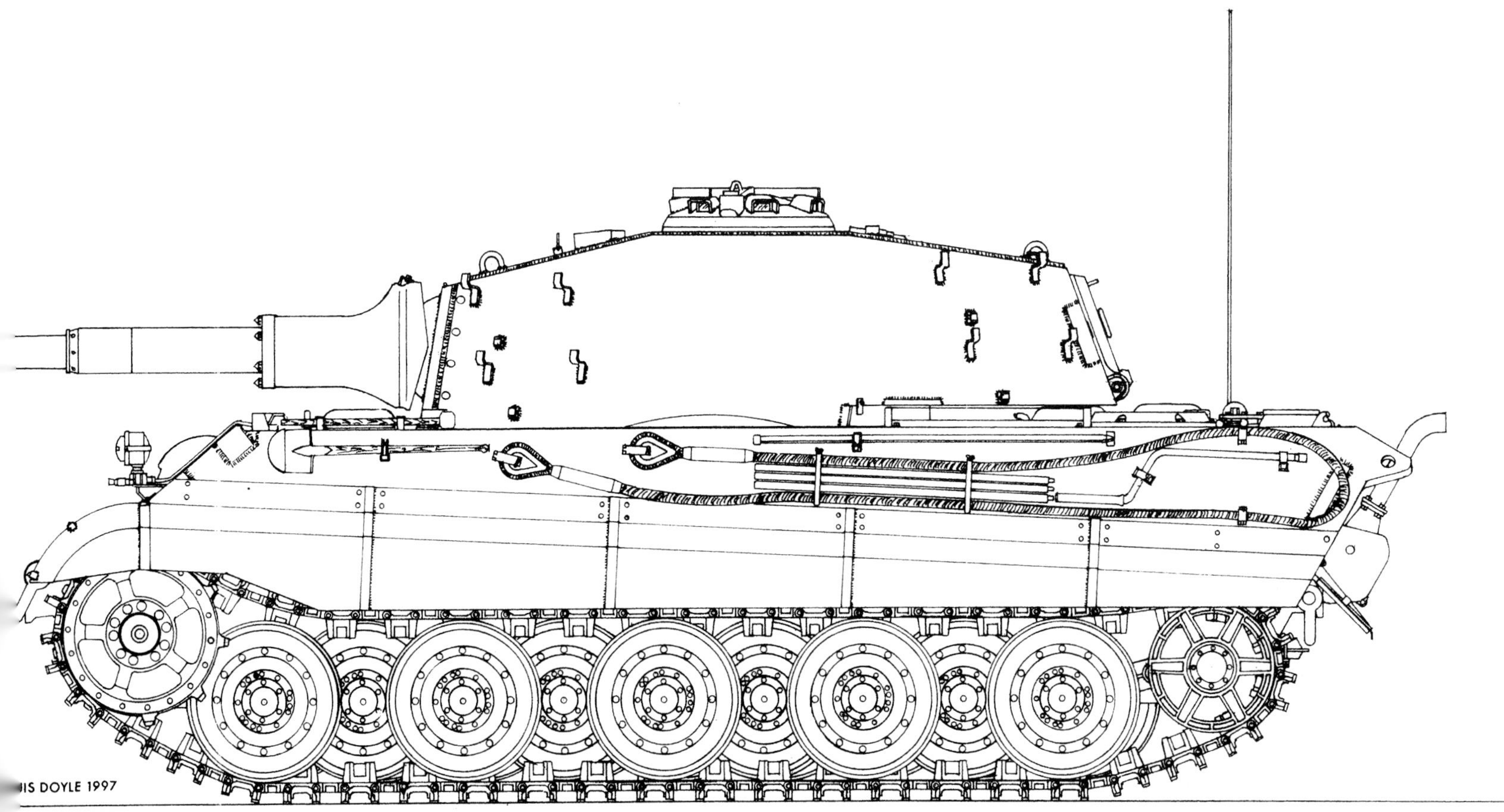
IS DOYLE 1997

Tiger Ausf.B (**Fgst.Nr.280243**, **Turm Nr.280091**) was completed by Henschel on 11 September 1944. It was first issued to **schwere Panzer-Abteilung 509**, who subsequently turned it over to **schwere SS-Panzer-Abteilung 501**. These size-comparison shots with a 76 mm Sherman were taken after it was captured by the Americans during the Battle of the Bulge. As can be seen, the Sherman is taller and therefore presented an easier target. (NA)

Tiger Ausf.B (**Fgst.Nr.280243**, completed by Henschel on 11 September 1944) on display at the Aberdeen Proving Ground prior to cutting out the right side of the turret and superstructure for display inside the museum. It was completed directly after the order was issued to stop applying **Zimmerit** and did not receive a base coat of **Dunkelgelb** (RAL 7028) paint. An armor cap was bolted over the vent for the engine compartment on the rear deck. (APG)

LEFT: The left front corner of **Turm Nr.280091** on **Tiger Ausf.B** (**Fgst.Nr.280243**) with the **12-Uhr** azimuth indicator, broken **T.Z.F.9d** hanging from the front bracket, and the gunner's hand wheel for traversing the turret. (NA) RIGHT: The gunner's position in **Turm Nr.280091** on **Tiger Ausf.B** (**Fgst.Nr.280243**) with the hand wheel for traversing the turret to the front, the hand wheel for elevating the gun (to the right under the gun), the hand lever for power traverse (under the seat), and the commander's foot rest behind the gunner's back rest. (NA)

LEFT: The mount for the **M.G.34** in the right front corner of **Turm Nr.280091** on **Tiger Ausf.B** (**Fgst.Nr.280243**). Details of the belted ammunition feed tray, foot-operated firing linkage, and spent cartridge chute can be clearly seen. (NA)
RIGHT: The safety switch on the right side of the gun, the loader's hand wheel for the auxiliary traverse, and the boot and lower housing for the gun elevating gear are all seen in this view of the inside of **Turm Nr.280091** on **Tiger Ausf.B** (**Fgst.Nr.280243**). (NA)

The ammunition stowage racks and the rollers to aid in loading the gun are seen in this view of the left rear corner of **Turm Nr.280091** on **Tiger Ausf.B** (**Fgst.Nr.280243**). Three gun cleaning rods had been thrown into the turret. (NA)

The rear of **Turm Nr.280091** on **Tiger Ausf.B** (**Fgst.Nr.280243**) with the rear hatch open, suspended on the torsion bars protected by armor guards. The sheet metal guard inside the rear hatch was pivoted forward. (NA)

Details of **Tiger Ausf.B** (**Fgst.Nr.280243**, **Turm Nr.280091**), now on display at the Patton Museum in Ft. Knox, Kentucky. (Disregard the camouflage pattern.) (TLJ)

Bases for mounting the **Losterkennungstafeln** (poison gas identification panels) were welded onto the top of the gun mantlet and onto the rear of the turret roof.

The **Huelsenauswurflochdeckel** (spent cartridge ejection port) on the turret roof.

The commander's cupola, countersunk into the roof and held by seven bolted arc segments underneath the turret roof.

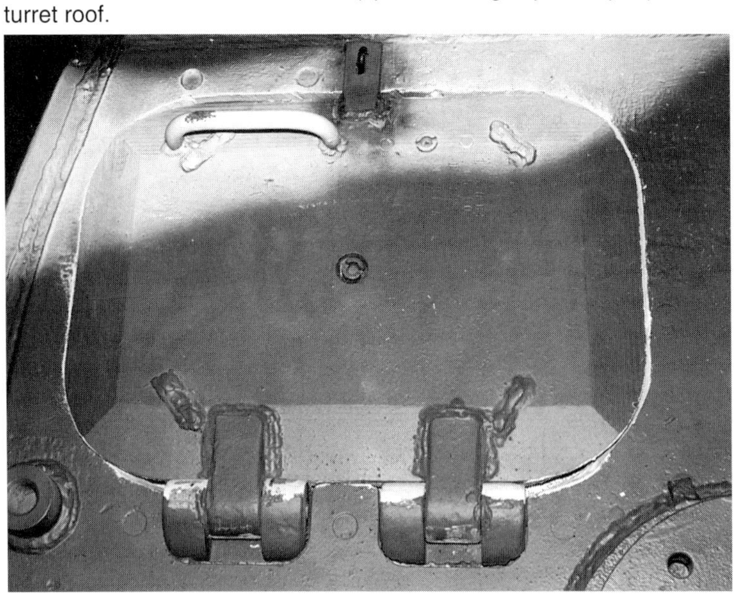

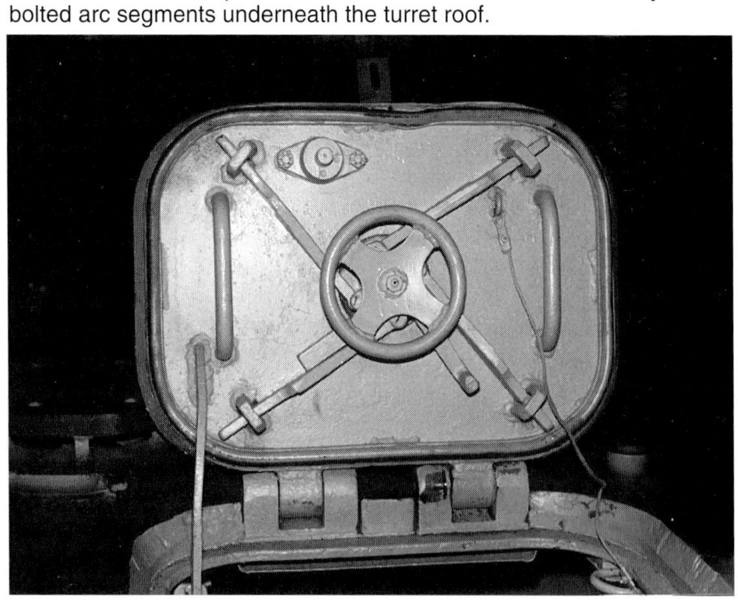

The 40 mm thick loader's hatch lid had two handles on the underside. Externally, both locks could be operated with a slotted key. The forked latch, used to hold the lid cracked open for ventilation, is missing.

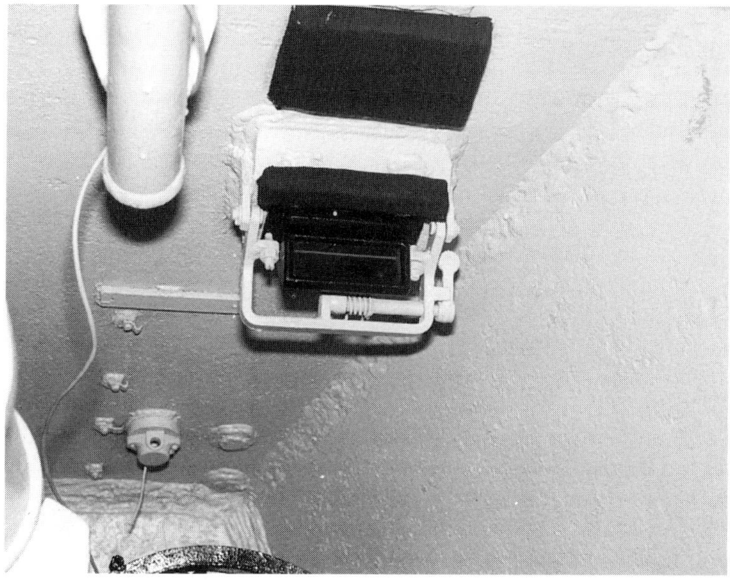

The loader's periscope in its mounting, which was designed so that a damaged periscope could be easily replaced, even with heavy gloves on in the winter.

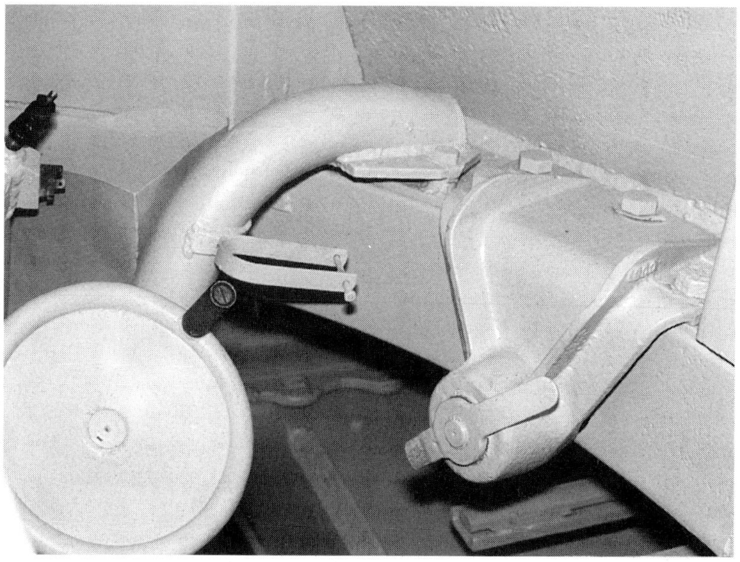

The auxiliary traverse handwheel for the loader and the turret traverse lock. The bar handle was easier to spin (than the previously used knob) for quickly disengaging the turret traverse lock.

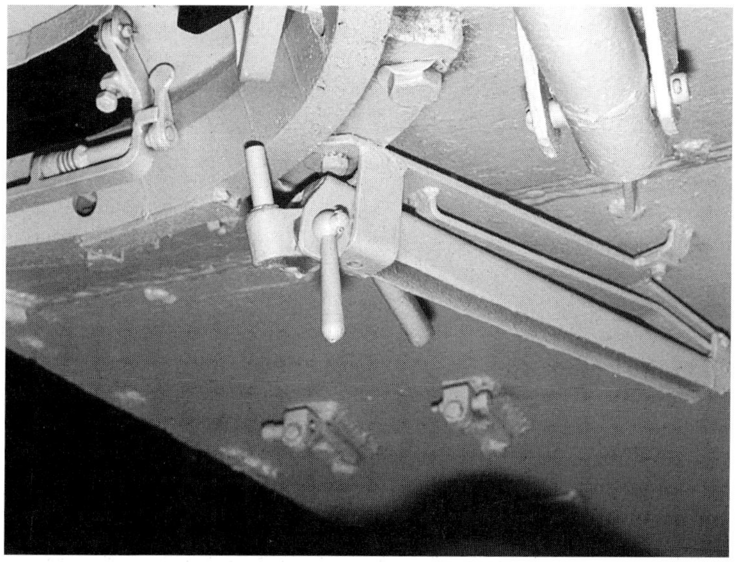

The sliding holder with head designed to mount the **S.F.14Z** (scissors periscope) for observation with the cupola hatch open. The quick-release periscope mounts were also installed in the commander's cupola.

The flexible base for the radio aerial countersunk into the rear deck. The triangular lid covered an opening for access to the wiring connections.

The wire screen is missing from over the radiator cooling air intake grate, but its sheet metal frame is still in place. The tube at the left rear, extending over the hull side, was a vent for the fuel system.

The roof of the turret on **Tiger Ausf.B** (**Fgst.Nr.280273**, completed by Henschel in October 1944 and issued to the **schwere SS-Panzer-Abteilung 501**) on display in La Gleize, Belgium. It has the commander's cupola bolted on and the 40 mm thick loader's hatch cover. As is usual for museum holdings, the cover is missing from over the turret ventilation fan guard. (HLD)

The rear deck of **Tiger Ausf.B** (**Fgst.Nr.280273**, completed by Henschel in October 1944) on display in La Gleize, Belgium. It has the armor guard bolted over the vent for the engine compartment and the armor caps in place over the fuel and water filler caps. The hole in the strip of armor behind the rear air intake grate was a penetration for a fuel system vent line. (HLD)

The rear of a **Tiger Ausf.B** completed by Henschel in October or November 1944 and issued to **schwere SS-Panzer-Abteilung 501**. The hangers for mounting the jack were no longer welded onto the lower rear. The tube for venting the fuel system penetrated the rear deck plate, ran down over the side, and continued forward ending just behind the rear holder for the wrecking bar. (NA)

The rear deck of **Tiger Ausf.B** (**Fgst.Nr.280404**, completed by Henschel in January 1945). As an experiment, sheet metal guards were mounted over the cooling air intake grates in order to prevent damage to the radiators caused by bullets from strafing aircraft or shell fragments. The inner edge of the forward pair of guards was cut out for turret traverse clearance. All four fuel system vent lines are visible in these photos, one at each rear corner and one each in front of the forward air intake grate. (WJS)

130mm

This **Tiger Ausf.B** completed by Henschel early in 1945 has the extended rain guard welded onto the turret front over the gun sight aperture. It is unique in having the first type of **Gg 24/800/300 Gelaendekette** (cross-country tracks) hung as spare track links on the turret sides. (WJS)

Tiger Ausf.B (**Fgst.Nr.280009** or **280012**) was found at Haustenbeck at the end of the war with single link **Kgs 73/800/152** cross-country tracks for a **Wa Pruef 6** test program. After the war, the British removed the **Kgs 73/800/152** tracks from this **Tiger Ausf.B** and mounted them on **Tiger Ausf.B** (**Fgst.Nr.V2**) now on display at The Tank Museum in Bovington, England. (TTM)

Closeup of the single link **Kgs 73/800/152** track with an 18-teeth drive sprocket on **Tiger Ausf.B** (**Fgst.Nr.V2**) on display at The Tank Museum in Bovington, England. (TLJ)

One of the last **Tiger Ausf.B**, completed by Henschel in March 1945, had the front mud guards with reinforced ribs and the single link **Kgs 73/800/ 152** tracks driven by 18-teeth drive sprockets. Pairs of single link tracks were also hung on the turret sides as spare track links, and the ring for mounting the anti-aircraft machinegun was still welded to the cupola periscope guards.

THIS PAGE AND OPPOSITE: These five completed turrets, ready to be mounted on **Tiger Ausf.B** chassis, were still at Henschel when occupied by American forces in late March 1945. All still have hangers for double link track welded to the turret sides. The last turret was outfitted as a **Panzerbefehlswagen Tiger Ausf.B** with an antenna base mount on the turret roof behind the loader's hatch. The ring mount for the anti-aircraft machinegun was not welded onto the cupola periscope guards on the second turret from the left. The anti-aircraft machinegun ring mount was present on the other four turrets. (NA)

THIS PAGE AND OPPOSITE: One of the last turrets for the **Tiger Ausf.B** completed by Wegmann. The abbreviation "**gepr.**" (inspected) and the date **21./3.** (21 March 1945) were chalked on the turret side. All of the last modifications had been completed on this turret. These included welding three sets of hangers fore and aft on both turret sides for spare single link track, welding five loops on both turret sides for holding camouflage, and welding a single post on top of the left front periscope guard on the cupola as the base for a swiveling anti-aircraft machinegun mount. This turret and ten bare armor hulls were shipped back to Aberdeen Proving Ground. (1-NA, 2-APG)

Panzerkampfwagen Tiger Ausf.B with a turret completed by Wegmann on 21 March 1945 – four vent lines for fuel system – dropped the jack, wooden block, and track mounting cable – rain guard over sight aperture – two locks for driver's and radio operator's hatches – ribbed track guards on front – single link **Kgs 73/800/152** tracks with 18 teeth sprocket – post mount for anti-aircraft machinegun instead of ring mount – five rings welded to each turret side for securing camouflage – spare track hangers for single link tracks on turret sides – simplified hatch on turret rear.

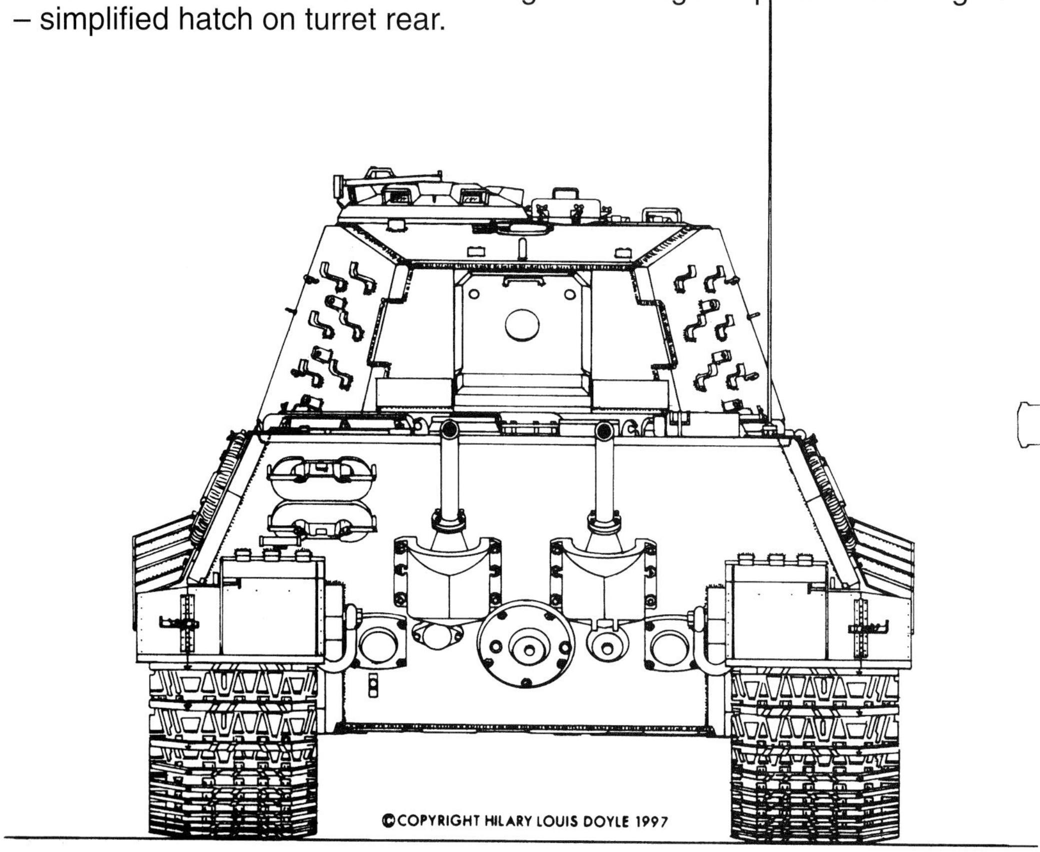

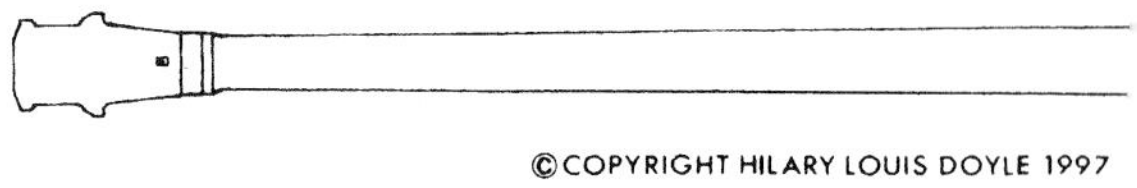

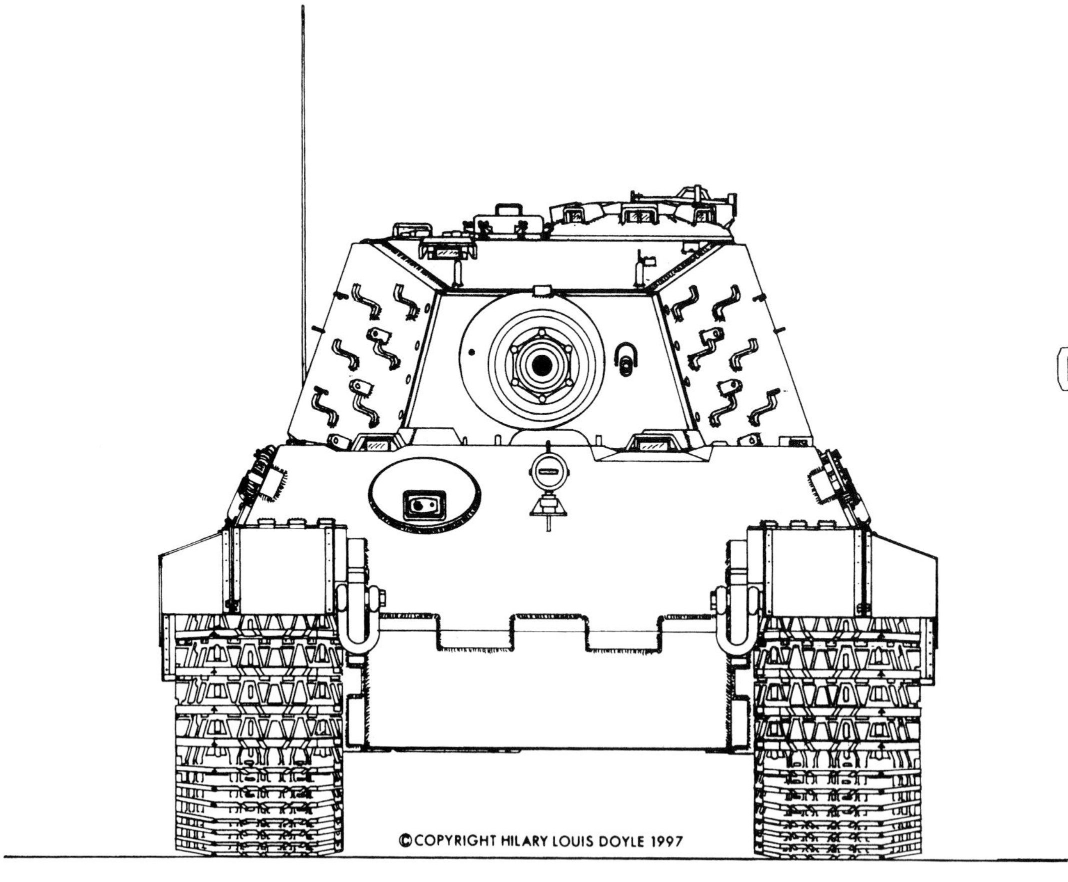

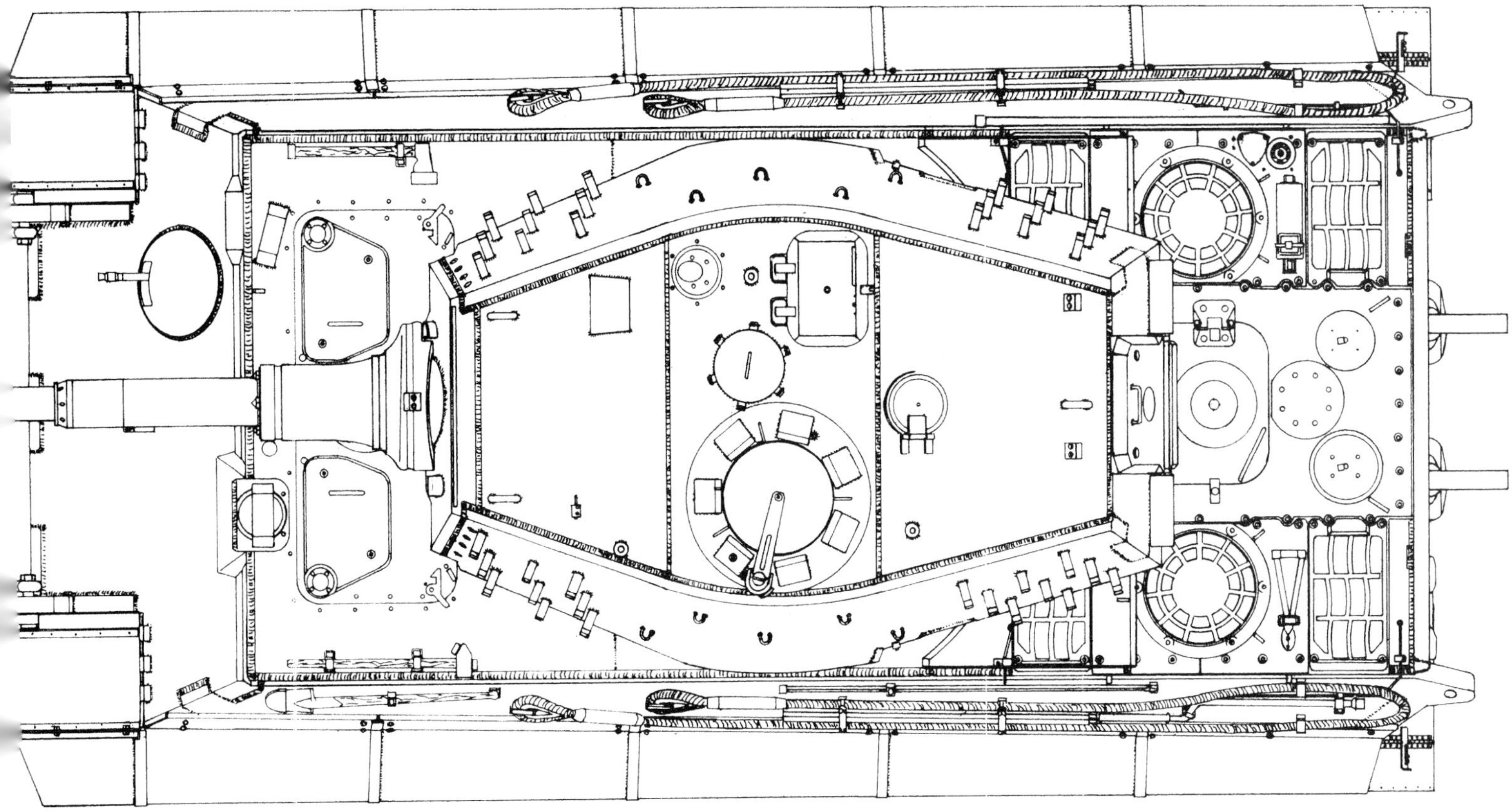

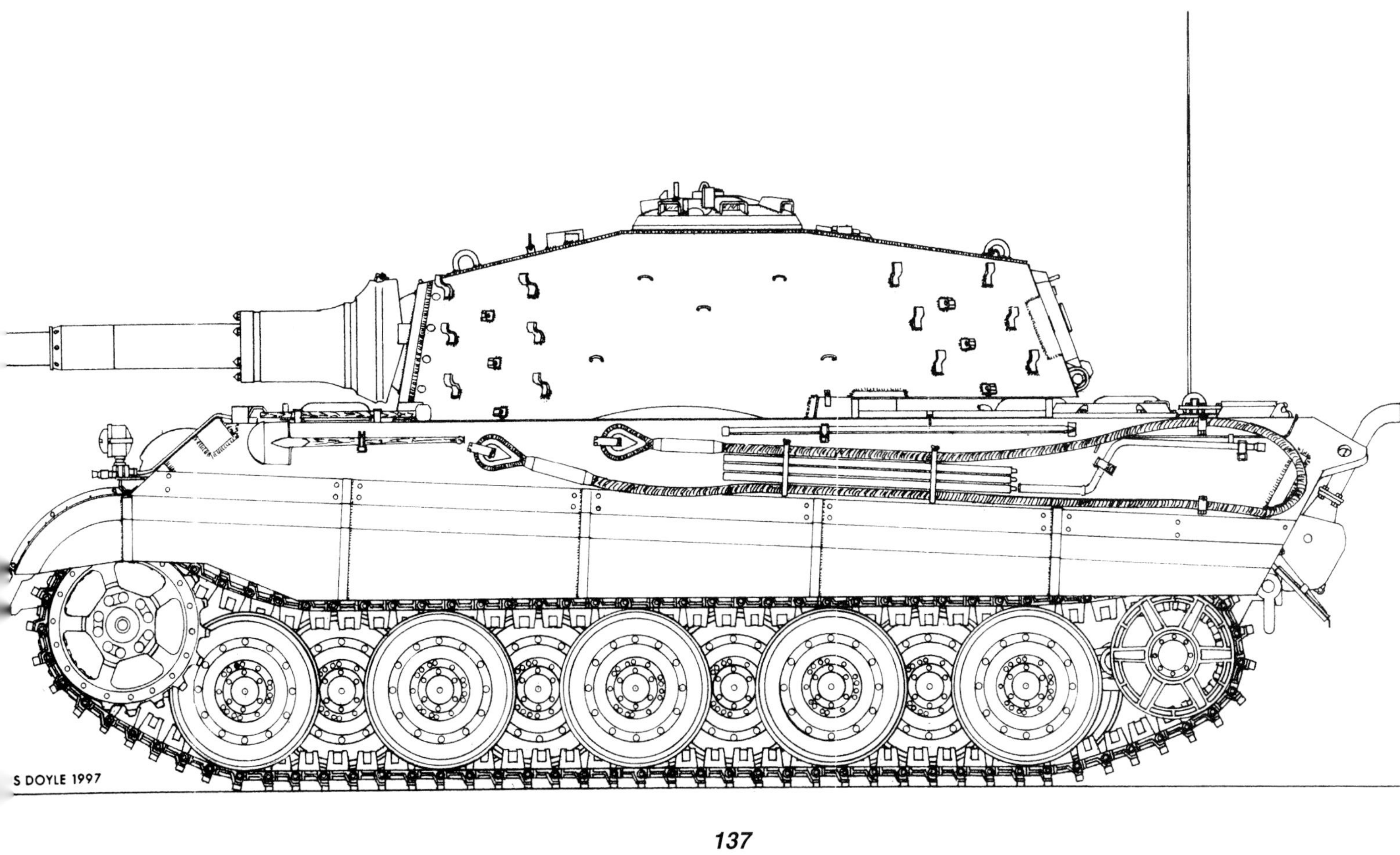

S DOYLE 1997

6.3.4 CAMOUFLAGE PAINT AND ZIMMERIT

Camouflage paint schemes on the Tiger II went through the following four phases:

1. Originally, the Tiger II left Henschel completely covered with a base coat of **Dunkelgelb** (RAL 7028) paint onto which the troops were to apply camouflage patterns using **Deckpasten** in **Olivgruen** (RAL 6003) and **Rotbraun** (RAL 8017) colors.
2. Starting after 19 August 1944, Tiger II were spray painted with the three tone camouflage scheme at the Henschel assembly plant prior to being shipped to the ordnance depot for issue to the troops.
3. Starting after 9 September 1944, **Zimmerit** and the base coat of **Dunkelgelb** (RAL 7028) was no longer applied. The camouflage scheme using **Dunkelgelb** (RAL 7028) and **Olivgruen** (RAL 6003) was sprayed sparingly onto the red oxide primer in which the armor components had been coated by the armor suppliers prior to delivery to Henschel and Wegmann.
4. Starting effective 1 March 1945, the assembly firms were to complete the **Buntfarbenanstrich** (multicolored) camouflage scheme by spraying on **Rotbraun** (RAL 8017) or **Dunkelgelb** (RAL 7028) in sharp contours onto the armor components which the armor suppliers were to deliver already covered with a base coat of **Dunkelgruen** (RAL 6003) paint.

6.3.4.1 Tarnanstrich (Ambush Camouflage)

On 19 August 1944, **OKH Wa J Rue (WuG 6) VIII** issued orders to all assembly firms: *Effective immediately, all armored vehicles are to receive a coat of camouflage paint at the assembly plant. In addition to a base coat of **Dunkelgelb** (RAL 7028), the colors **Olivgruen** (RAL 6003) and **Rotbraun** (RAL 8017) must be applied in patches. This concern is of decisive significance for the troops. Every effort must be made to complete part of the August production with this new camouflage pattern.*

This was the order that initiated what has become known as the "ambush" camouflage pattern. Prior to this, all Tiger II had been delivered to the troops with a base coat **Dunkelgelb** (RAL 7028) paint and each individual unit had applied its own camouflage pattern.

6.3.4.2 Leuchtfarbenanstrich (Luminous Paint)

On 25 August 1944, as stated in a decree from the **OKH Wa J Rue (WuG 6) VIIId**, effective immediately **Panzerkampfwagen** and **Sturmgeschuetz** were no longer to be painted on the inside with **Leuchtfarbenanstrich** (luminous paint). Henschel stopped painting the inside of the Tiger II with **Leuchtfarbenanstrich** starting with **Fgst.Nr. 280177**, completed about 25 August 1944.

6.3.4.3 Zimmerit (Anti-Magnetic Coating)

On 9 September 1944, **Wa Pruef 6 PzIIIb** informed the assembly firms: *Based on a decision by the **Generalinspekteur der Panzertruppen** on 7 September 1944, for the purpose of simplification, effective immediately, the protective coating (**Zimmerit**) against **Hafthohlladungen** (magnetic shaped charges) is no longer to be applied to armored vehicles.*

6.3.4.4 Base Coat of Red Oxide Primer

As specified by the **Heeres-Abnahmestelle II(K), Hannover** on 31 October 1944:

*Immediately stop painting the inside of all **Pz.Kpfw.** The hull and components are to remain in the condition in which they were delivered. The assembly firms are no longer to paint the inside of the **Panzerkampfwagen**. Painting the outside is to accomplished by applying patches of **Braun, Gruen, und Gelb Deckpast** (RAL 8017, RAL 6003, and RAL 7028) onto the hulls delivered in red oxide primer. If **Gelb** isn't available, in extreme emergencies **Graue** (RAL 7021) paint can be used; otherwise **Graue** paint is to be conserved.*

This directive, issued to MNH in Hannover in late October, must have been issued previously to Henschel in Kassel close to the same time that **Zimmerit** was ordered to be dropped on 9 September 1944. Tiger II (**Fgst.Nr. 280243**), completed on 11 September 1944, left the Henschel assembly plant with only a base coat of red oxide primer (RAL 8012) onto which wide, irregular stripes of **Olivgruen** (RAL 6003) and **Dunkelgelb** (RAL 7028) were sparingly sprayed.

6.3.4.5 Base Coat of Dunkelgruen (RAL 6003)

On 29 November 1944, further orders were issued by **OKH Wa J Rue (WuG 6) VIII** regarding a new camouflage paint scheme. The external surfaces of all armor components were to be delivered to the assembly plants already covered with a base coat of **Dunkelgruen** (RAL 6003) paint. The assembly firms were then to complete the **Buntfarbenanstrich** (multicolored) camouflage scheme by spraying on **Rotbraun** (RAL 8017) or **Dunkelgelb** (RAL 7028) in sharp contours. The new camouflage scheme for completed Panzers was to go into effect starting on 1 March 1945. By 20 December 1944, armor suppliers had been issued orders to immediately start coating armor components with **Dunkelgruen** (RAL 6003) paint prior to delivery.

On 23 January 1945, a further directive on the subject of painting **Panzerkampfwagen** was issued by **OKH Wa J Rue (WuG 6) VIII** as follows:

After further extensive investigation, and based on requests from the troops and companies, the following is established to expand on the previous order dated 29 November 1944:

*1. It is permitted and again ordered that the interior of the turrets (or by **Sturmgeschuetz** the surface area above the track covers) be painted with a coat of **Elfenbein** (RAL 1001) paint.*

2. So long as available paint stocks are adequate without thinning until conversion to the new camouflage paint directive occurs starting on 1 March 1945, the old paint stocks can continue to be used up to 30 May 1945.

*3. In order to make it possible to achieve standardization in the camouflage paint scheme, the companies are to inform Direktor Fuchs of the **Fachabteilung Lackbedarf des Heeres im Produktionsausschuss Lacke beim Reichsmin.f.R.u.K.** of the paint supplies that they need.*

4. Of course, as far as it is possible, companies should already convert to the new camouflage paint scheme prior to 1 June 1945.

As is evident in a color photograph, at least one Tiger II completed by Henschel in March 1945 was completely covered with a base coat of **Dunkelgruen** (RAL 6003) paint upon which thin, wandering stripes and spots of **Dunkelgelb** (RAL 7028) paint had been sprayed to create the camouflage pattern.

Tiger Ausf.B (**Fgst.Nr.280006**), completed by Henschel in February 1944, was issued to the **Panzerversuchsstelle** in Kummersdorf. As with all Tigers produced up to mid-August 1944, it left the factory with a base coat of **Dunkelgelb** (RAL 7028) paint applied over the entire outer surface, including the **Zimmerit**. (BA)

Varying camouflage patterns were applied by the unit, as is evident in photos like this one of a **Tiger Ausf.B** completed in June 1944 and issued to the **3.Kompanie/schwere Panzer-Abteilung**. Stripes of **Olivgruen** (RAL 6003) and **Rotbraun** (RAL 8017) had been spray painted over the base coat of **Dunkelgelb** (RAL 7028). (BA)

As shown in this photo of a **Tiger Ausf.B**, completed by Henschel in early July 1944, the maintenance platoon was equipped with an air compressor and was responsible for spray painting the Panzers with camouflage patterns that would blend in with the local terrain. (BA)

This **Tiger Ausf.B** (**Fgst.Nr.280354**, completed by Henschel on or about 19 December 1944) was painted at the assembly plant with **Dunkelgelb** (RAL 7028) and **Olivgruen** (RAL 6003) spray painted in dots and patches only partially covering the base coat of red oxide primer. All **Tiger Ausf.B** were to be issued a **Regenschutz** (canvas rain guard), but it is rare to see it erected over the commander's cupola. (BA)

Even when the camouflage pattern was sprayed on at the assembly plant, there was no systematic pattern. Compare this pattern on **Tiger Ausf.B** (**Fgst.Nr.280356**) completed at Henschel on or about 20 December 1944) with that on **Tiger Ausf.B** (**Fgst.Nr.280354**) completed at nearly the same time. These two **Tiger Ausf.B** being towed around by **Jagdtiger** (**Fgst.Nr.305004**) took part in an experiment for determining the proper location of towing hooks needed for field recovery. (BA)

One of the last Tigers, produced by Henschel in March 1945, was completely covered with a base coat of **Dunkelgruen** (RAL 6003) paint over which **Dunkelgelb** (RAL 7028) paint was sprayed to create a camouflage pattern of irregular stripes and dots.

6.4 MODIFICATIONS AFTER ISSUE

The **Waffenamt** authorized the field units to implement a series of improvements on their Tiger II after issue. These included:

a. Starting in June 1944, weld three **Pilze** for **Befehlskran 2t** on the turret roof.

b. Starting in July 1944, replace the 18-tooth with a 9-tooth drive sprocket ring (or cut off every second tooth) and fit the new double-link **Gelaendeketten** with its single-piece connecting link.

c. Starting in October 1944, field units were ordered to stop applying **Zimmerit** anti-magnetic coating to Panzers.

d. Starting in November 1944, fabricate and bolt a cap over the **Lufteintritt** (vent for the engine compartment) on the rear deck.

e. Starting in December 1944, weld track hangers onto the turret sides for **Ersatzkettenglieder**.

As reported on 28 February 1945, **Wa Pruef 6** experimented with a Tiger II (**Fgst.Nr. 280404**) by covering the air intake gratings on the rear deck with thin steel plates to prevent damage to the radiators caused by bullets from strafing aircraft or shell fragments. There is no evidence that this modification was authorized for implementation by Henschel in the production series or by the troops after issue.

The troops themselves did not make many unauthorized modifications. One of the infractions, evident from photographs, was hanging additional spare track links from the turret and superstructure sides. For some reason, infrequently they braced the tail pipes with a spare track link or a track guard section.

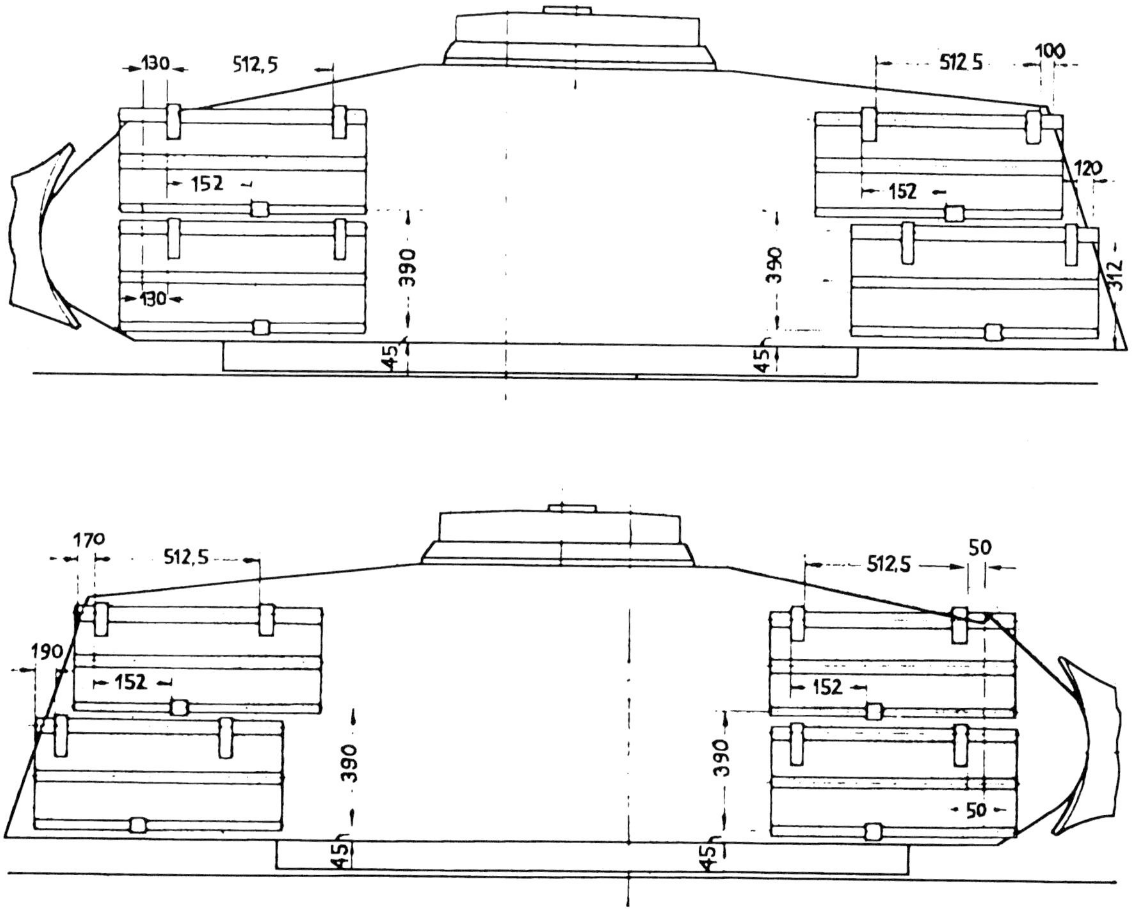

The list of modifications to the **Tiger Ausf.B**, published in the H.T.V.Bl. on 15 December 1944, included instructions on welding spare track-link hangers to the sides of **Turm Nr.1-50** as shown in the drawings. Instructions had also been included for backfitting these hangers to **Serienturm** produced in June and early July.

6.5 MODIFICATIONS UNDER DEVELOPMENT

The war came to a close before a significant number of unique state-of-the-art modifications could be completed and incorporated into the production series. Henschel alone was working on 19 modification packages for improving the chassis. Krupp and the firms of Zeiss and Leitz were busy working on improvements to the weapons and fire control, including a built-in range finder, stabilized sights, and a stabilized gun with automatic ammunition feed. Other unique ideas, such as a rounded turret for the **8.8 cm Kw.K.42 L/71**, an **M.G.151/20** in the turret for anti-aircraft defense, and a **10.5 cm Kw.K. L/68** in the Tiger II turret, were abandoned at the conceptual design stage.

6.5.1 SUMMARY OF FUTURE MODIFICATIONS

On 6 February 1945, Dr. Aders, director of the design department at Henschel, recorded the following status of the design projects for the Tiger B chassis:

In addition to and after working out the bugs to create a mature design, there was still much work to be accomplished on further development and to take advantage of field experience.

1. Install a new gasoline engine from Maybach-Motorenbau, the HL 234 with fuel injection that could replace the previous HL 230 without any additional work to make it fit.

2. Install various diesel engines, of which two (Simmering-Pauker-Graz and Argus-M.A.N.) were air cooled. Both of these engines required an entirely new engine compartment layout and very considerable changes to the hull armor and to the track tensioner. A third engine from Humboldt-Deutz is water cooled and less promising.

*3. As desired by the **Heereswaffenamt**, cooling the transmission by an air current drawn through a sheet-metal housing turned out to be insufficient, just like in the Panther. Cooling the circulating lubricating oil in the water-cooled system, proposed by us six years ago and turned down as unnecessary, again had to be taken up and designed by modifying the cooling system.*

4. A torque-limiting, slip clutch on the radiator fan designed by Sueddeutsche Argus-Werke was simplified and given to the shop for testing.

5. To employ the vehicle in very cold combat zones, various proposals were made for heating the fighting compartment. The last one was tested and approved for series production.

6. The air intake gratings on the rear deck redesigned for better protection against hits (rolled bars were curved to improve air flow and welded inside a cast frame) were again redesigned as castings. The results are questionable and can first be proven after driving trials using a wooden model.

7. In the event that the Maybach transmission could no longer be produced or shouldn't prove itself suitable, installation of the transmission, designed by Zahnradfabrik Friedrichshafen for the Panther, was investigated and preparations made.

*8. Installation of a transmission of a special design (Puls-Leipzig) that was contracted by the **Heereswaffenamt** was investigated and preparations made for driving trials in a Tiger E.*

*9. Installation of a hydraulic steering system (without stages, with continuously changing turning radius), contracted by the **Heereswaffenamt** from Zahnradfabrik Augsburg, was investigated and preparations made.*

*10. The design of the final drive shafts (between the brakes and the steering unit) was modified based on experience. The requirement from the **Heereswaffenamt** for standardization that it also had to fit into the Panther II had resulted in a complicated and less robust design.*

11. The attempt with electrical-welded idler wheel that was introduced for other vehicles is also to be exploited for simplifying production of the Tiger B. Preparations were being made for extensive trials.

*12. The safety device in the track tensioner, dropped from the Tiger E as requested by the **Heereswaffenamt**, is to be installed again and immediately tested. One piece of the adjusting element will be designed as the weakest point so that when it breaks because of a shock, other more valuable components will be protected.*

*13. The deliberations over the design of a flexible track tensioner did not lead to usable results. As proposed by the **Heereswaffenamt**, we worked on a hydraulic tensioner with cylinders and pistons, in which balance between both tracks and common cushioning was provided by an air cushion.*

14. Even more so than those on the Tiger E, the newest track demonstrated the disturbing characteristic that the track pins drift out toward the hull, catching and breaking on the rear edge of the hull. Prevention against this is inadequate. A deflecting and returning device has to be obtained and tested.

*15. As a result of experience by the troops with other types of Panzers, the **Heereswaffenamt** wants to test a auxiliary device that can be mounted on the track drive wheel. If the final drive fails and jams, this device should make it possible for the track to slip around the drive sprocket and thus make it possible to tow the vehicle.*

16. Destruction of ball-bearing industry factories led to replacing ball-bearings with roller-bearings. In the area of the Panzer production, in the interim the measures were proven to be impractical and technically not producible. However, replacement parts were designed and tested for the Tiger E and the Tiger B.

17. The rear deck above the engine compartment, which we were directed to take over from the Panther design, turned out to be unsuitable, was redesigned, and is to be incorporated into production starting at about the 700th Tiger B. To decrease the weight of the frequently used hatches, the hatch cover is divided into three parts and the size of the opening has been increased for easier access to the engine.

18. Experience in the field by the troops showed it to be necessary to install two towing rings on the front plate and one centered on the rear in addition to those that were cut into the hull side extensions. No fewer than 13 proposals were necessary for the rear centered tow coupling. Then only after practical trials were conducted was approval given for acceptance into the production series.

*19. There were significant deficiencies in the driver's and radio operator's hatch covers designed to be lifted and then pivoted to the side (Henschel was required to copy this design after those on the Panther). Temporarily the **Heereswaffenamt** held the opinion that these needed to be replaced by hatch covers that are spring balanced and can be jettisoned in an emergency. After laborious design work and testing, the requirement apparently has been dropped.*

20. As a preventive measure against future poison gas attacks, a protective device for installation in the Tiger B and Panther was designed by the especially suitable firm of Draegerwerk-

Luebeck . The device (consisting of a filter box, transfer case, and a fan to overpressurize the fighting compartment) took up a lot of space and was difficult to install. At this time installation is planned, and operation in the Tiger, Jagdtiger, and Panther is being tested in Henschel's proving grounds under very considerable expenditures of equipment.

21. As a new item, it is planned to secure the gun tube in a travel lock for rail travel and for long road marches without enemy activity. Work is in progress.

22. Two large projects are mentioned that aren't directly connected with the development of the ***Panzerkampfwagen****: Use of the chassis without turret as a* ***Schuttraeumgeraet*** *(designed after a snow plow) to clear off streets after bombing attacks and the development of a* ***Gleisketten-Schleppers mit Dampfbetrieb*** *(fully tracked tractor with a steam engine). The first device is now under construction for the city of Kassel; the other device is still being designed.*

On 20 February 1945, the **Chef des Heeresstabes Ruest** sent the approved lists for the **Entwicklungs-Notprogramm** (development emergency program) to the **Wa Chef Pruef**. The top priority list consisted of inventions that had a decisive influence on conducting current battles, could be implemented in the near term, and had to be accomplished as the core development program. Development of new modifications on the top priority list associated with the Tiger II were expected to be completed in:

April 1945 - **Stabilisierung der Optik** (stabilized gun sights)
April 1945 - **Entfernungsmesser** (rangefinder)
August 1945 - **Maybach-Motor HL 234 (900 PS)** (fuel-injected engine)

These are the dates that the designs were to have been completed, not the later dates when the modification would be ready to enter series production. Other design projects such a diesel engines were not sufficiently advanced for inclusion in the top priority list.

A second priority list was included that consisted of inventions on which work couldn't be totally stopped without significant loss of research and development work, continuation of which would not have an impact on the battlefield in the near term but whose development promised special advantages in the future. Modifications for the Tiger included on this second list were:

Luftgekuehlter Simmering-Dieselmotor (air cooled diesel engine)
Wassergekuehlter Deutz-Dieselmotor (water cooled diesel engine)
Stabilisierung der Waffe (stabilized main gun)
Gasschutz Sondereinbauten in Panzer (poison gas protection installed in the tank).

6.5.2 FUTURE DRIVE TRAIN COMPONENTS

On 6 September 1944, Maybach sent **Wa Pruef 6** a report on the status of development of the Panzer engines which included the following details on the **Maybach HL 234 Motor**:

The ***HL 234*** *is a 12-cylinder V-engine with 23 liter swept volume. Its expected power is at least 800 horsepower at 3000 rpm. The engine works by directly injecting fuel into the cylinders. With this it is possible to reduce the fuel consumption at peak load and especially under partial load when compared to the* ***HL 230****.*

The cylinder head gasket will consist of an iron ring along with individual thick rubber rings at locations where the cooling water and oil could enter the cylinder head from the engine block. [The head gasket for the **HL 230** was made of copper and was not reusable. The compression ratio for the **HL 230** had to be decreased and the engine speed limited to 2500 rpm.]

The high power ratio for each cylinder requires especially good cooling of the combustion chamber, which will be achieved by new cooling water flow paths and enlarged water pumps. [The **HL 230** had problems due to the close proximity of the back-to-back cylinder liners and insufficient cooling]

A strong high-shoulder bearing, located outside the oscillation damper, serves to horizontally stabilize the crankshaft. Another high-shoulder bearing is installed outside the flywheel so that the crankshaft turns in nine bearings. [The cause of engine breakdown in **HL 230** engines was often failure of one of the seven crankshaft bearings].

The oscillation damper and the flywheel are pulled inside the engine block. This results in small crankshaft penetrations on the end of the motor which can be easily sealed against oil leaks.

A normal electric motor serves as the starter. The inertia starter is organically built into the engine.

The underside of the engine is completely flat.

The connections for the ***HL 234*** *are designed so that it can be installed in the* ***Panther I G*** *and in the* ***Tiger B*** *with only changes of the type that will still allow the use of the* ***HL 230*** *as a replacement engine with no additional changes.*

Plans were made to deliver the first ***Versuchsmotor*** *to* ***Verskraft Kummersdorf*** *(proving grounds in Kummersdorf) in early 1945.*

The following report on a meeting of the **Entwicklungskommission Panzer** under **Die Reichsminister fuer Ruestung und Kriegsproduktion** held on 23 January 1945 provides details on the engine development program for **Panzerkampfwagen**:

In opening, Dr. von Heydekampf referred to the following guidelines: The special requirements are 1) engines with higher power, 2) conversion to diesel engines, and 3) air-cooled diesel engines.

Based on the state of developments, advancing step-wise is the correct approach as reflected by the listed priority of the special requirements. In the name of the ***Generalinspekteur der Panzertruppen****, General-Major Thomale demanded these advances. He requested more powerful engines for the Tiger and Panther and added that, with regard to engines, the soldiers already "have been burned too often." He reiterated the history of problems with the* ***HL 230*** *and demanded that only healthy engines with really higher horsepower be delivered to the troops. From experience in the last few years, he warned against hastening to deliver insufficiently tested and debugged engines to the troops. Otherwise, as occurred before with the acceptance of the* ***HL 230****, hundreds of Panzers will stand around unusable. One can now no longer afford to let the men bleed due to deficient engines in the Panzers and to throw away the expensive armor in the Panzers due to this cause.*

a. Wassergekuehlter Maybach-Otto-Motor HL 234

*Oberst Holzhaeuer (**Wa Pruef 6)** stated that two years ago the carburetor of the **HL 230** had already been recognized as a weakness and attempts to convert to fuel injection had been initiated. The Maybach **HL 234** had been created from these developments and, according to the latest reports from Maybach, achieved 900 metric horsepower. Fully loaded down, fuel consumption was 220 grams/metric horsepower; partially loaded, somewhat improved over the **HL 230**. Also, the head gaskets, connecting rod bearings, and main crankshaft bearings had been improved. In all details, the **HL 234** engine could replace the **HL 230** engine. Dr. von Heydekampf injected that a report on the trials was not yet available and that the **HL 234** engine had not been tested in a **Panzerkampfwagen**. Because of the stated advantages, the decision was unanimous that the **HL 234** be immediately included in the engine design and procurement program.*

b. Luftgekuehlter Simmeringer Diesel-Motor

Oberst Holzhaeuer stated that development of the Simmeringer Diesel-Motor was started more than two years ago at the request of Hitler. Trials on the test stand have been satisfactorily concluded. Eight cylinders of this 16-cylinder engine lay down in the hull and are poorly accessible. When installed in an armor hull it was shown that the cooling air fan was not satisfactory. A new fan may be available in about four weeks and trials in a Panzer can be accomplished. The production expenditure for the Simmering engine is significantly higher than the Maybach engine, about double.

*Installation in the Panzer requires minor modification of the hull. The Simmering engine is not exchangeable with the Maybach engine. The actual state of development allows the start of a **0-Serie** of a total of 100 engines that are to be produced by Steyr.*

*Direktor Ochel added that a rapid increase in the production of Simmering engines is planned. Their production capacity freed up from the Luftwaffe should be used for the Panzer sector. Steyr can't convert their series production twice, once to the Tatra engine and then to the Simmering engine. Dr. von Heydekampf's opinion is that such an increase in Simmering engine production is not justifiable. The **Entwicklungskommission Panzer** has to urgently warn that untested engines are being hastily shoved into mass production. General-Major Thomale and Oberst Holzhaeuer emphatically agreed with this statement. It was decided that Oberst Holzhaeuer and Dr. von Heydekampf present this opinion in a memorandum in a meeting with Reichsminister Speer.*

c. Luftgekuehlter Argus-MAN-Diesel-Motor

The development of this engine is about one-quarter year behind the Simmering engine. There is unanimous agreement that the Argus-MAN engine should remain in development until it has been shown that the Simmering engine is proven successful in a Panzer.

d. Wassergekuehlter Kloeckner-Humboldt-Deutz-Diesel-Motor.

*Development started on this engine about one year later than both air-cooled engines after it was recognized that the air-cooled engines wouldn't be ready for production as quickly as had been expected. The Deutz engine is a V-8-cylinder motor of sturdy design. All eight cylinders are accessible in the Panzer, it works as a two-stroke engine, and according to Dr. Flatz it can achieve a power output of 800 horsepower. It has good torque characteristics and is simple to produce. It is suitable for installation in a Panzer as a reliable diesel engine with the same space requirements as the current **Maybach HL 230**, and therefore development should be accelerated. Direktor Ochel's report on the actual shortage of development capacity has to be urgently reported to Kloeckner-Humboldt-Deutz.*

e. AK 7-200 in Tiger II

*It is possible to install the **AK 7-200** transmission in the Tiger II. According to a report from **Wa Pruef 6**, driving characteristics are not as good as the **Olvar-B** transmission. Direct exchange of the transmissions are not possible. Installation, investigation, and testing are still not completed. Plans remain to continue production of the Tiger II and the Jagdtiger with the **Olvar-B** transmission.*

6.5.3 CREW COMPARTMENT HEATER

As mentioned by Dr. Aders, a crew compartment heater had been designed and approved for series production. As shown in drawing number 021 B 49534 dated 25 October 1944, air heated by the radiators on the right side was diverted into ducts which fed the warm air into the crew compartment. This design required modification of the rear deck around the right side fan louver. Because the modification order allowed the armor suppliers to complete already fabricated parts, it is very doubtful whether this modification was incorporated into any of the production series Tiger II completed by Henschel through March 1945.

6.5.4 AMMUNITION STOWAGE

On 20 November 1944, Henschel completed a sketch for a new proposal to increase the amount of main gun ammunition stowed in the **Tiger Ausf.B**. Along each side in the panniers, eight rounds were to be stored in the front, nine rounds in the center, and thirteen rounds in the rear (instead of the previous six, seven, and eleven). This would have increased ammunition stowage in the hull by 12 rounds.

In addition to the racks in the side panniers, Henschel completed drawing HSK 4372 dated 12 February 1945 for a **Zusatzmunitionshaelterung** (additional ammunition rack) for five rounds to be stowed horizontally behind the turret platform in front of the firewall. Further detailed drawings of the racks needed for the expanded ammunition stowage scheme in the panniers were dated February 1945, too late for acquisition and installation before Tiger II production ceased in March 1945.

6.5.5 STABILIZED MAIN GUN AND AUTOMATIC AMMUNITION FEED

During a meeting with **Wa Pruef 6** on 24 January 1945, Krupp was informed that the company of Kreiselgeraete A.G. in Berlin had been awarded a contract for the development of a stabilizer for a tank gun because this company had extensive experience in the area of gun stabilization. However, this did not lock Krupp out from also receiving a contract. Krupp had already reached an agreement with Oberst Crohn that Fried.Krupp would begin conceptual design work without a contract on a stabilizer for a long gun tube.

On 20 February 1945, Krupp was awarded a contract from **Wa Pruef 6** to design a Tiger II turret with a **vollstabilisierte, starr eingelagertem Geschuetzrohr 8.8 cm L/71** (fully stabilized, rigidly mounted 8.8 cm L/71 gun tube). At the same time, conceptual design of a **voll- oder halbautomatische Munitionszufuhr** (fully or partially automatic ammunition feed) was to begin.

An excess turret located at Wegmann was to be converted for trials. The impact of the recoil on the turret ball-bearing race was to be especially studied and a stabilizer proposed by Kreiselgeraete A.G. in Berlin was to be tested. The Krupp proposal for a gun stabilizer, assisted by an offsetting mounted weight instead of a gyro, was to be considered further.

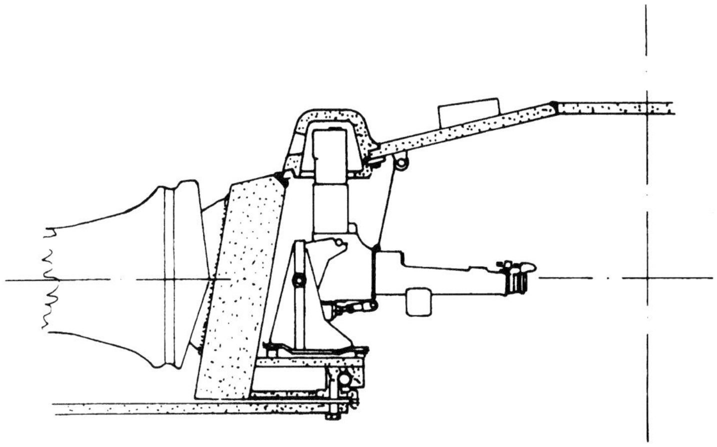

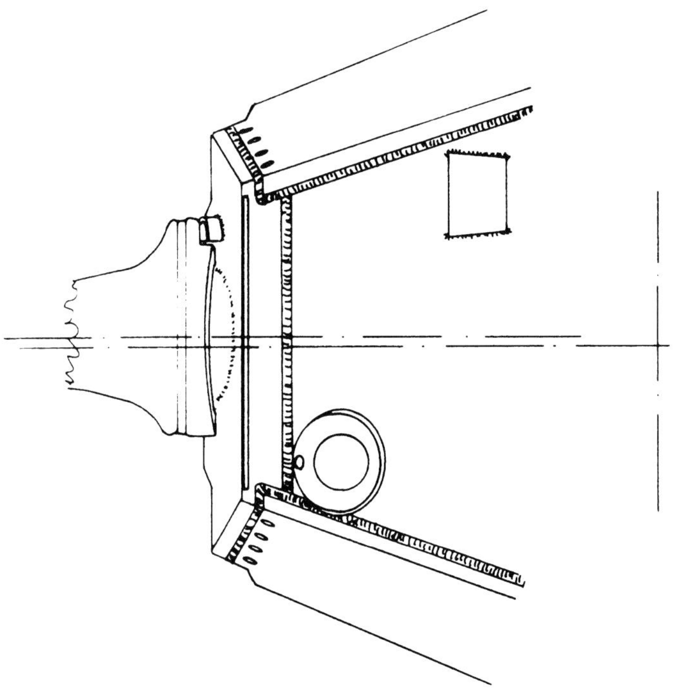

ABOVE: A preliminary sketch of the mounting for the **Turmwinkelzielfernrohr 3** (stabilized, periscopic gun sight) installed in a **Tiger II** turret from drawing Hln-A 130 dated 23 October 1944.

BELOW: A crew compartment heater diverted air heated by the radiators on the right side into ducts, which fed the warm air into the crew compartment.

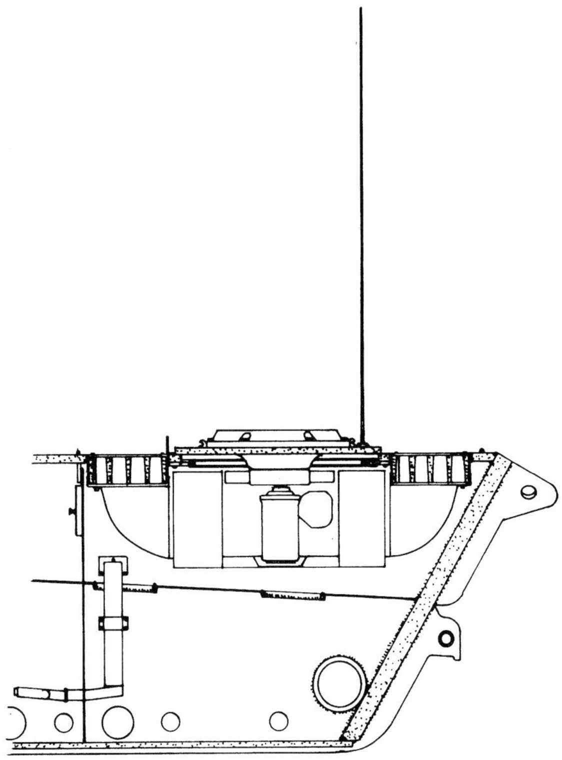

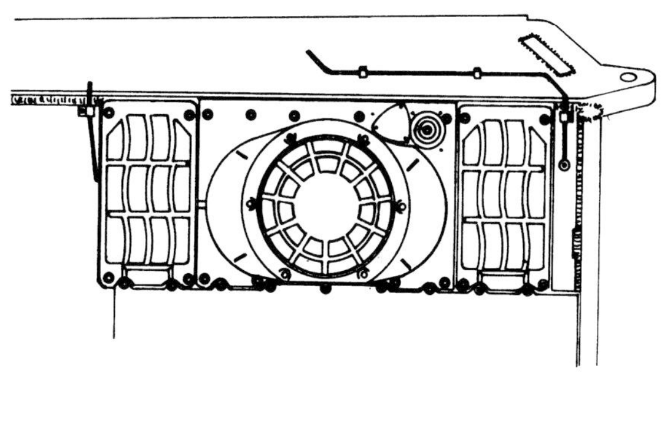

6.5.6 MOUNTING AN M.G.42 IN THE TURRET

On 8 August 1942, **Wa Pruef 6** wrote to Krupp about designing a mounting for an **M.G.42** in the turret: *A new cushioned mount for the **M.G.42** must be developed. Whether or not the previous belted ammunition sack holder for machinegun ammunition can be retained can first be decided after viewing the design of the mount. It is intended that the previously accepted belted ammunition sack be used and to feed the belt through a metal tube the same as in aircraft. In connection with this, determine if a container for 2000 rounds of machinegun ammunition can be installed in the turret.*

In a meeting on 19 February 1943, **Wa Pruef 6** met with representatives from Krupp, Rheinmetall, and Daimler-Benz to discuss the proposed designs from all three firms for mounting the **M.G.42** in Panzers. Krupp and Rheinmetall were to examine their proposed designs and strive to take over as many parts from the Daimler-Benz design as possible in order to create a standard machinegun mount. The fastener for this standardized mount was to be left to the individual firms for fitting into their respective turret designs. Krupp and Rheinmetall were to present modified designs by 10 March 1943. The design firms were ordered to keep the hole in the gun mantlet as small as possible and to install a deflector shield on the inside to catch small caliber bullets or shell splinters. It was determined that the **M.G.42** in its current shape could be installed without modification into the **Panzerkampfwagen Panther und Tiger**.

[Records have not been found revealing why the **M.G.42** had not already been installed in 1943. In January 1944, **schwere Panzer-Abteilung 506** requested that they be provided with an **M.G.42** mounting in the turret as the **M.G.34** still suffered from too many stoppages and that they could carry an adequate supply of ammunition for the higher rate of fire from the **M.G.42**. An armor sleeve had been designed for the **M.G.34** for protecting the protruding barrel when it was mounted in the Panzers.]

It was only after the decision was made to halt **M.G.34** production that the next mention of installing the **M.G.42** in the Tiger occurred during a meeting on 21 November 1944. It was decided that starting after March 1945, one **M.G.42** was to be mounted in the turret and a **Sturmgewehr 44** was to be mounted in the ball mount for the radio operator in the **Pz.Kpfw.Tiger**. The **M.P.40** was to be retained as the crew weapon. [Since Tiger II production ceased in March 1945 and other production delays had occurred, it is very doubtful that a single Tiger II was produced with an **M.G.42** in the turret.]

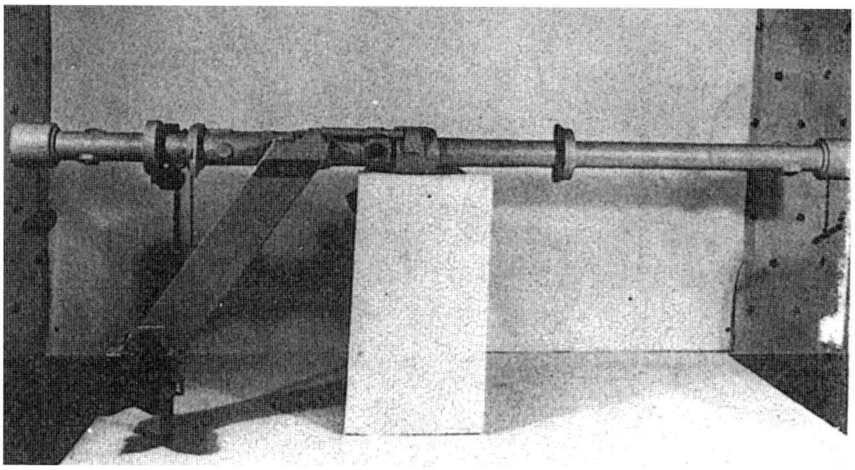

The **Entfernungsmesser 1.6 m R (Pz.)** (1.6 meter wide base range finder) designed to be mounted inside a modified Tiger II turret. (NA)

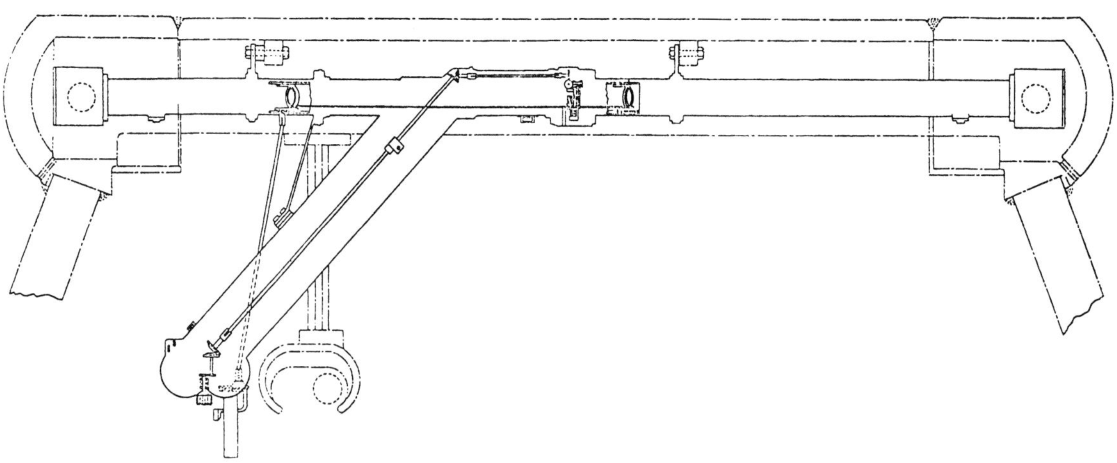

A drawing of the **Em. 1.6 m. R (Pz)** range finder designed to be mounted inside a modified **Tiger II** turret. This cutaway view shows the mechanism for turning a prism so that the images would overlap in this stereoscopic range finder. The degree to which the prism was turned was displayed on a scale as the range to the target.

6.5.7 STABILIZED GUN SIGHTS

Details on the design basis and capabilities of stabilized sights were reported as follows by Herr Ernst Haass inventor and designer, by Kreiselgeraete A.G., Berlin (incorporated into CIOS Report No. XXXII-34 dated 27 August 1945):

While moving, it is desirable for self-propelled guns to pursue a target once it has been sighted and if possible to fire at it while in motion. This task is opposed by considerable difficulties. When driving over more or less rough ground, the angular height of the gun and the sight fitted to it change continuously. The target appears sometimes above, and sometimes below, the hairline. This is very strenuous and tiring for the observer. At long ranges, a high magnification factor is required for the sight. However, this high magnification factor further increases the difficulties in observing a target which dances up and down in the cross-hairs. A magnification factor of two is, therefore, generally used for non-stabilized sights. However, this is totally insufficient for long range fire. Thus, if observation of the target is very difficult while in motion, a successful engagement is impossible while in motion. This difficulty can be eliminated when the gun barrel along with the sight is stabilized by a gyro mechanism. This has been achieved with good success. However, this direct stabilization is suitable only for smaller guns up to 37 mm caliber. For larger guns, very big gyros must be used on account of the sharply increasing moment of inertia of barrel, recoil, and imbalance of the barrel. This has the disadvantages of being heavy, taking up a large space, requiring a lot of power, and being costly to manufacture. These factors cannot be justified by the vehicle. For large caliber guns the only way left open is to stabilize the field-of-view.

The field-of-view stabilizer can be built very small, light, accurate, and cheap, because it works practically without a load. Such instruments were built, but at first without a pre-ignition gyro. This meant that observation of the target was possible without difficulty while the vehicle was moving. The target rests quietly in the cross-hair and can be observed with any magnification factor devised, even at a range of 10,000 meters. However, to fire, it was necessary to stop the vehicle to make the center of the barrel coincide with the sight. Installation of the pre-ignition gyro represents a further improvement. It's task is to measure the barrel's angular speed and to effect a correction that corresponds to the firing retardation. This mechanism permits firing and hitting while the vehicle is in motion.

The instrument has to fulfill two functions:

a. The exit prism of a telescope that is fitted to the cradle of a gun barrel has to be steered in such a manner that it does not follow the changes in angular height that occur when driving over uneven ground. Neither is it to follow the changes in angular height from the sighting axis directed toward the target that occur when the barrel traverses up and down. The gunner has to keep the target caught in the cross-hairs and maintain it there in spite of upward and downward movements of barrel and telescope.

b. The instrument is to fire the round by electric contact when the gun barrel traverses upward and downward just at the moment at which the barrel is at the correct angular height for the target. A potential mistake, which may be caused by the firing retardation and angular speed of the barrel, can be eliminated by a correction mechanism that is dependent on the angular speed.

Results: At the first trials the instrument was tested for many hours on a test track. Faultless observation was possible without the observer tiring in the least. The instrument is simple and sturdy and works without the least failure. As a result of the well-arranged gyros, it is completely insensitive to all movements of the gun or vehicle beyond its elevation.

Test firing of 10 rounds resulted in a mean value deviation of _0.5 meters from the target at 1000 meters distance, corresponding to an angular value of 0.5 mils. [This test firing was conducted with a modified telescopic gunsight with gyrostabilizer attached as **S.Z.F.1** mounted in a Panther with the **7.5 cm Kw.K.42 L/70** gun.]

On 10 March 1945, Krupp reported to **Wa Pruef 6** that they had tried to fit the **S.Z.F.3** into a Tiger B turret with a range finder. The sight, including the gyro drive, could be mounted in the turret in the same location and use the same hole as the previous telescope. The turret front plate needed to be modified to change the size and shape of the external aperture.

Ernst Leitz, Wetzlar reported that they had received contracts to produce 1001 **S.Z.F.3** as order number 5702001. However, they did not report that any of them had been produced or delivered.

The firm of Carl Zeiss, Jena, had also worked on the development of a stabilized, periscopic gun sight, as they reported postwar on 19 June 1945:

The need to observe and shoot while on the move led to a periscopic gun sight design with gyro-stabilized view. The design was made in such a manner that as few changes as possible had to take place in the manufacturing of the sight with or without gyro stabilization. Drawing No. 335545 is the **Turmwinkelzielfernrohr 2** without stabilization and drawing No. 335547 with stabilization. The stabilization worked in such a way that the objective lens and the sight reticle were stabilized. Gun movement and elevation were also connected to the sight reticle. The gun could be fired when the sighting mark covered the target. Firing could occur only while stationary; observation while on the move; side deflections were not stabilized. For firing on the move, a pre-ignition device with automatic firing was chosen which took into consideration the high angular speed of the gun barrel.

The **Turmwinkelzielfernrohr 2** and **3** were still being designed at the end of the war. Acceptance was delayed because it wasn't possible to achieve the required time for manufacturing.

6.5.8 RANGE FINDER MOUNTED IN THE TURRET

The war in North Africa had shown that range measurement in such a country would be most useful, as guessing the range was misleading. When the tank halts before firing and the range to the target is under 1000 meters, the elevation of a high-velocity gun is not effected much by range, and range measurements can be dispensed with. The great advantage of using a range finder lies in the possibility of opening fire at long ranges. Without the range finder, precious time is lost when the elevation of the gun must be corrected by observing the effect of fire. As the ammunition supply in tanks is limited, each shot which does not hit is a double loss of time and radius of action. These disadvantages are eliminated by the use of a range finder. Therefore, designs were initiated for mounting a rangefinder inside a Panzer.

On 17 June 1943, representatives of Zeiss and Leitz met with Krupp and **Wa Pruef 6** to discuss the range finder design, as follows: *Using the available drawing 343159, Zeiss clarified how the **Versuchs-Em fuer Tiger** was to be installed and which work needed to be done on the Panzer. On 30 May, Krupp had re-*

Panzerkampfwagen Tiger Ausf.B with changes that should have entered production by July 1945 – raised turret roof for internal range finder – cooling air louvers on rear deck – cover for engine compartment vent on rear deck – three-piece hatch for motor compartment – crew compartment heated by diverting air warmed by radiators – **Sturmgewehr 44** instead of **M.G.34** in ball mount for radio operator – coax **M.G.42** instead of **M.G.34** in turret.

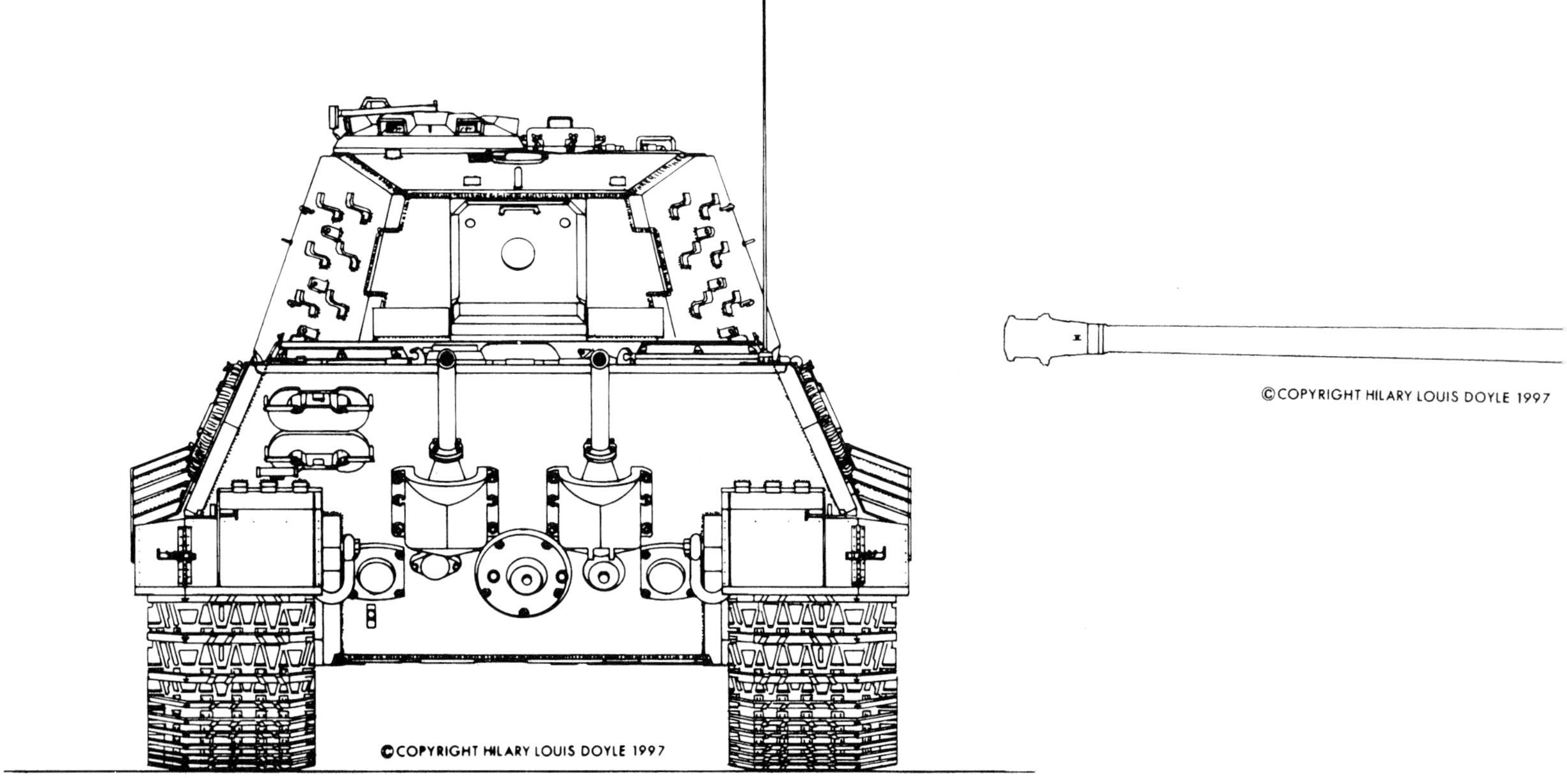

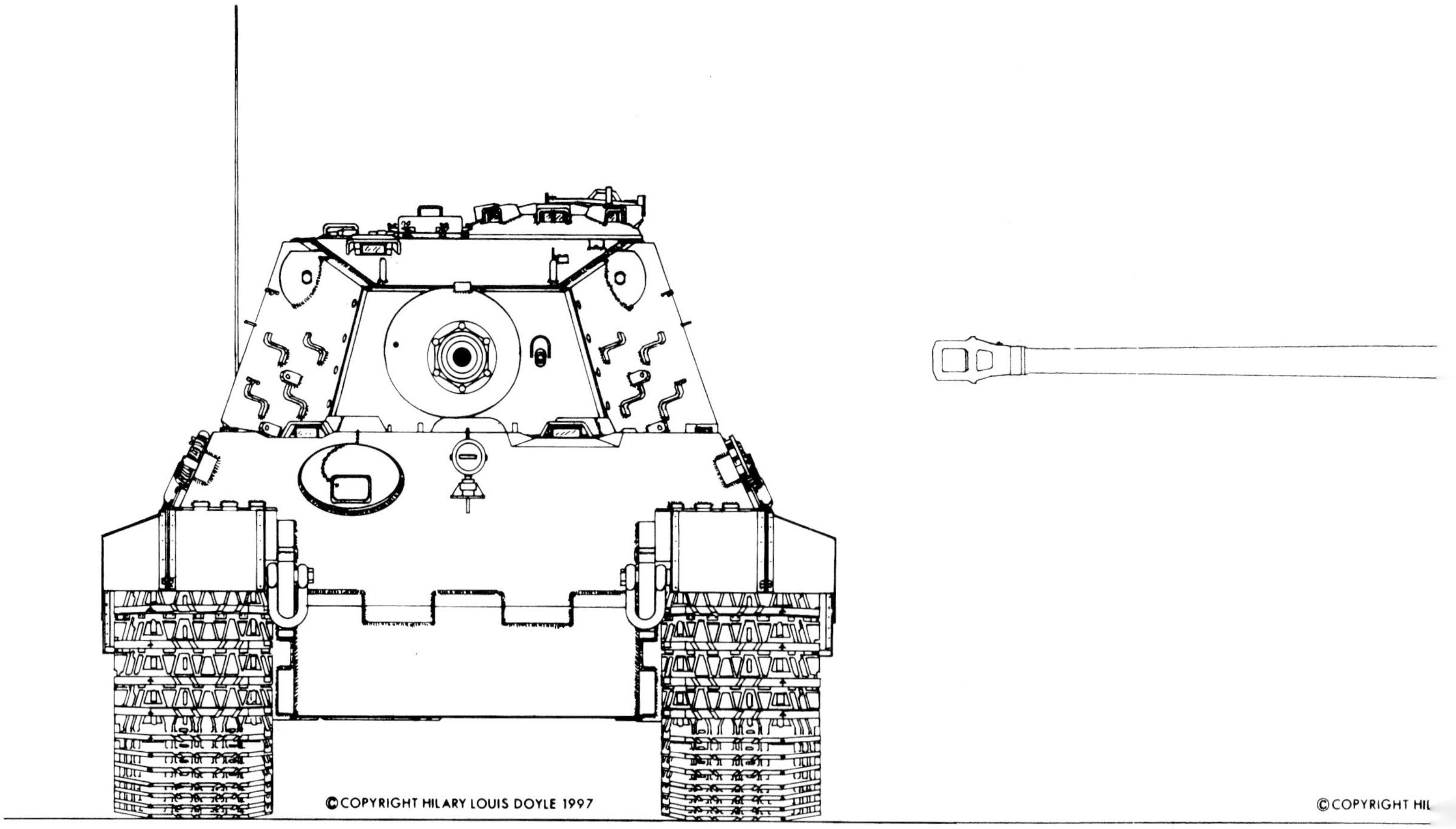

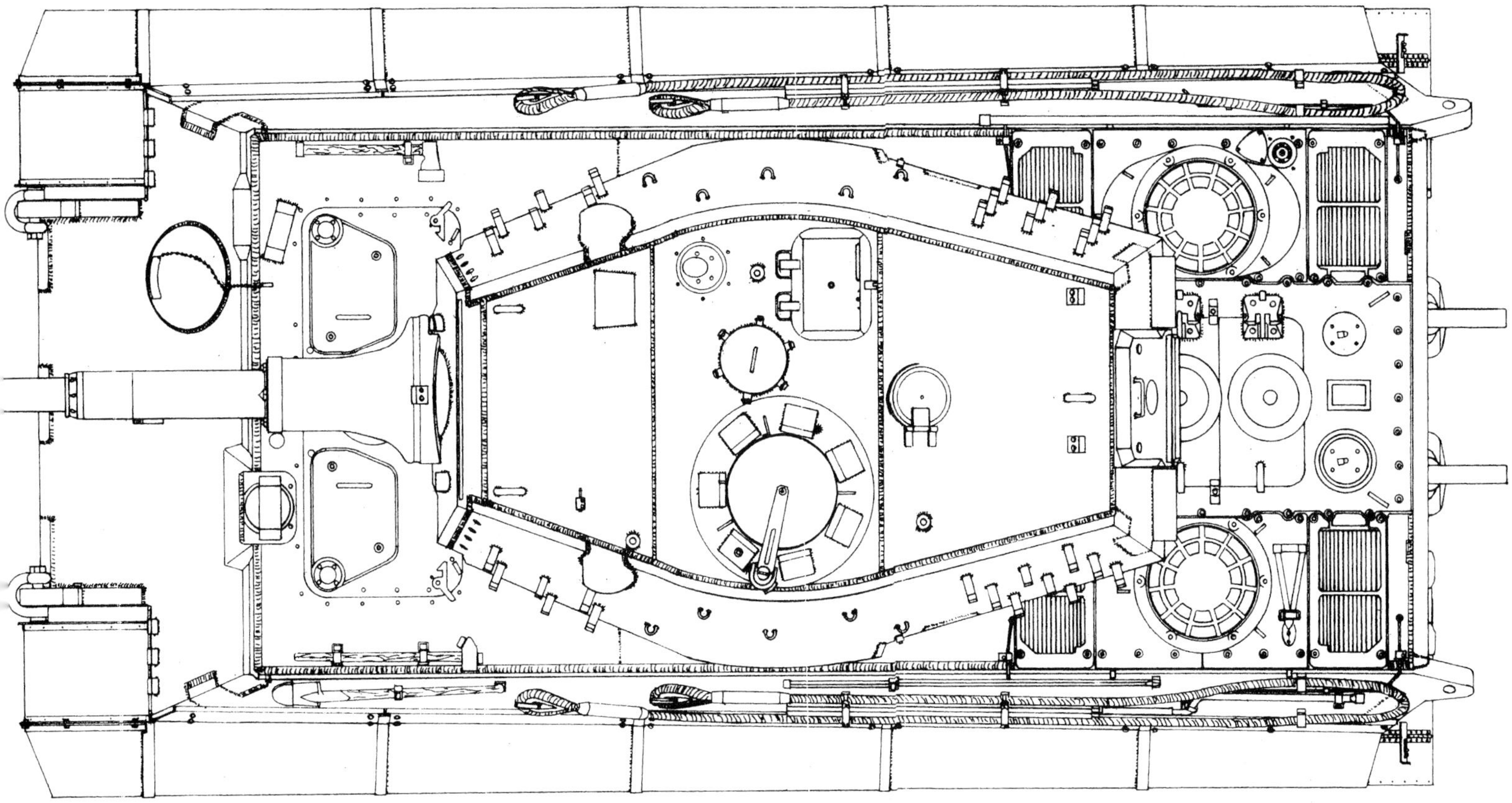

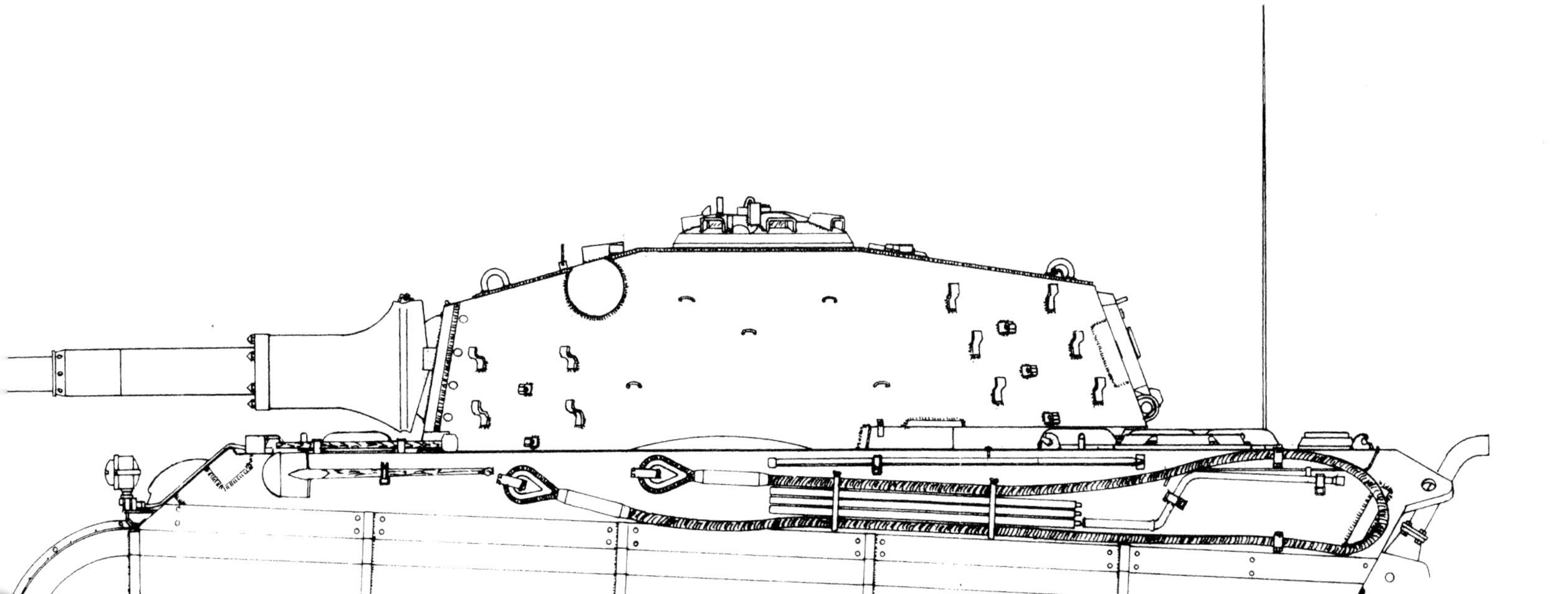
S DOYLE 1997

ceived from Zeiss a wooden model of the range finder which had been ordered on 22 April. Oberstlt. Crohn didn't consider that a rotating shield would be suitable, especially considering that dirt would eventually enter, and that the area between the Panzer and the rangefinder had to be completely sealed. In principle he liked the seal with an opening for cleaning, and it was to be constructed this way.

Directly after the War, Zeiss described their **Em. 1.6 m. R (Pz)** range finder designed for mounting in the Tiger II turret as follows:

This instrument is a 1.6 meter base stereoscopic range finder designed for mounting in the turret of the Tiger II. The instrument is suspended from the turret roof, stretches right across the turret, and the end windows are placed behind small ports at the front of the upper turret wall. An interesting feature is the periscopic arrangement by means of which an inclined tube carrying the eyepieces allows these to be positioned slightly to the left of, and at approximately the same level as, the eyepiece of the telescopic gun sight. The gunner can therefore use either the range finder or the telescopic gun sight with only a small movement of his head. The length overall is 168 cm, height overall 64 cm, and maximum width 26 cm.

Optical data: Magnification 15 x, diameter of exit pupil 2 mm, aperture 30 mm, field of view 4° (71 meters wide at a range of 1000 meters), optical height 46 cm.

As the tank commander is preoccupied with other tasks, the gunner must take charge of the range finder. After locating the target with the periscope and aligning the target in the sight, the gunner operates the range finder, reads the range, and adjusts the range scale to the correct range which provides the additional elevation associated with the range.

The range scale is seen at the bottom of the field of the right eyepiece. It is graduated in meters from 550 to 20,000 and marked as follows:

Ranges	Graduations	Numbered every
550-600 meters	5 meters	10 meters
600-900	10	20
900-1200	10	50
1200-1500	20	100
1500-2000	50	100
2000-3000	100	200
3000-4000	200	1000
4000-20000	500	10000

Beyond 6000 meters, ranges can only be read to about 500 meters, and the instrument is not designed for measuring such ranges.

Drawing 27-3240.001(C) shows the mounting of the 1.60 meter rangefinder for Tiger II. It needed the slightest changes to the turret of the normal tank. It was only necessary to set ears on the turret and to raise the roof of the turret to gain 25 mm in height for mounting the range finder. The ears are perforated by long slits as lookouts for the range finder at angles from -9 to +15°.

Tests were conducted to prove that decelerations of up to 12 g (associated with bottoming out after crossing a trench) would not disturb the adjustments of the range finder. After completion of a test course of 20 kilometers, measurements were repeated and all errors were found to lie within the limits of approximate uncertainty of the range finder. In the course of these experiments, it was also confirmed that the accelerations caused by firing the gun are minimal and may be neglected in comparison to the mechanical strain on the range finder in a moving tank.

Rubber blocks were installed as dampers to isolate the range finder from the turret in order to protect the range finder against damage or misalignment when the turret was hit by armor-piercing shells. Tests were conducted by firing AP shells of up to 7.5 cm at the turret armor, causing peak acceleration of 2200 g. But checks on an installed range finder proved that the rubber buffers were dampening the shock and not affecting its adjustment even though some shocks were measured at over 12 g. This satisfactory result was the more impressive, as one round had penetrated the armor and chips had damaged a screw sitting near the eyepiece. These experiments proved that the mounting of the range finder could be designed in a way to make it as invulnerable as any other instrument used in a tank.

The first results of shooting using a range finder mounted in a tank were not satisfactory, as the target was not hit. All shots were too high. Consequently a systematical error was recognized and the opinion was formed that the range finder had made a mistake.

A control made by measuring the range of the target with geodetical means proved that the range taken with the range finder lay within the limits of the allowable errors. The real cause of the poor shooting was found after systematic research showed that:

1. The aiming line of the range finder was not exactly adjusted to the axis of the gun. The coupling between the elevation axis of the gun and the vertical sliding movement of the mark in the range finder was not rigid enough and the connecting bar gave way.
2. The composition of the powder charge did not correspond to the ballistic data and the necessary combustion temperature in the gun was not maintained.

All these faults are inconsequential for shooting without range finders, because they are compensated for while adjusting fire, but they must be minimized when the range is taken exactly if good accuracy is to be obtained. These conditions, well known to field artillery, aren't sufficiently regarded in tank gunnery at the moment.

A series of 2000 range finders with mountings fabricated in accordance with drawing Nr. 27-3240.001 were contracted from Carl Zeiss, Jena. A total of three Em. **1.6 m. R (Pz)** were produced by Zeiss: one in November 1943, one in December, and one in February 1944.

But it wasn't as easy to install the range finder in a turret as Zeiss had inferred, as related in the following letter from Krupp on the **Tiger B Turm mit E-Messer** sent to Wa Pruef 6 on 13 January 1945:

*Following a meeting in Dortmund on 21 December 1944, DHHV is ready to rapidly produce the range finder turret. The final supporting documents needed for production were sent to DHHV by courier on 9 January. In several days, the drawings of the armor ears will be sent to **Wa Pruef 6**.*

*New difficulties have arisen because the **Ausblasezylinder** (pneumatic cylinder specified by **Wa Pruef 4** for blowing fumes out of the gun to replace the compressor mounted on the turret floor) only allows the gun to be depressed to 6° in the turret design which has already been completed for mounting the range finder. In order to depress the gun to the specification of 8°, the* *height of the turret armor body must be increased another 25 mm, which means a total of 45 mm in comparison to the* ***Serien-Turm****. The 45 mm increase brings with it an increase in weight of 110 kilograms. In addition, the commander's cupola must be moved inward toward the gun another 20 mm, leading to an unfavorable space situation.*

In the current ***Serien-Turm*** *being produced without a range finder, the* ***Ausblasezylinder*** *can be installed without modifying the turret and the specified gun depression can occur.*

Decisions made on 25 January 1945 to alleviate the problem were reported as follows: *Depression of the gun at 6° must be temporarily accepted in the Tiger II turret with range finder. Also, install a stop in the elevation mechanism that can be easily removed by the troops. Immediately investigate if the fumes can be blown out of the gun by using the front end of the recuperator cylinder. In addition to this, if occasion arises, plan on longer and smaller radius* ***Ausblasezylinder*** *so that gun depression of 8° can quickly be achieved again. A model Tiger turret with two range finders must be presented quickly. Report to* ***Wa Pruef 6*** *when the turret will be delivered.*

On 14 February 1945, the **OKH Waffeninspektion Ruest (WuG 6) VIIId** reported on the subject of the **Sattelbach Turm E-Messer**: *All Tiger B are to have a range finder built in, starting with the July 1945 production. Therefore, the turret armor bodies must be completed in accordance with the new drawings from Krupp.*

On 28 February 1945, Krupp reported to **Wa Pruef 6**: *DHHV is striving to complete their first turret armor body for housing a range finder by 31 March 1945. Conversion to the range finder turret at Krupp could be achieved starting with the 601st turret, which means about mid-July 1945.*

6.5.9 ABANDONED DESIGNS

In addition to modifications started too late to make it into production, Krupp and **Wa Pruef 6** started work on a series of conceptual designs that didn't make it off the drawing boards. These included a new round turret for the Tiger 2, mounting an **M.G.151/20** in the Tiger II turret for anti-aircraft defense, and mounting a 10.5 cm gun in the Tiger II turret.

6.5.9.1 New Tiger 2 Turret Design

On 18 September 1942, Krupp wrote to **Wa Pruef 6** that they had designed new turrets, as follows:

On 6 July 1942, during a meeting on Panzer turrets we gave ***Wa Pruef 6*** *proposals numbered 1 AKF 31460 and 4 AKF 31462 as well as 1 AKF 31435 and 1 AKF 31436 on turrets for 8.8 cm L/56 and 8.8 cm L/71 with forged armor bodies.*

In the interim we have followed this direction in design farther and are sending new proposals AKF 31510 and AKF 31511 for a "Tiger 2" turret with forged armor body in which the ***8.8 cm Kw.K.42*** *currently in production can be mounted without modification. Equally good, all of the important components such as the traversing and elevating mechanisms, turret platform, traverse mechanism support and drive, and turret ring can be practically taken over unmodified. The turret doesn't have a protrusion at the rear. Acceptable rate of fire is achieved by stowing the ammunition on the right side of the turret platform along with a lifting device that grabs and lifts the cartridge bases. Loading the gun can occur at all elevations.*

This turret is thought by us to be a follow-on type to the current "Tiger 2" in line with simplifying production of the turret armor body.

The weight of the complete turret is 11 metric tons, in comparison to 13 metric tons for the previous model. The commander's cupola is substantially simplified, because the armor ring housing the periscopes can be rounded. Earlier protection of the turret ring against hits has been reinstated by dropping the base ring that was needed to support the side walls on the welded design.

6.5.9.2 Installation of an M.G.151/20 in a Tiger II Turret

On 17 February 1943, Krupp representatives met with Oberstlt. Crohn (**Wa Pruef 6 II**) to discuss mounting anti-aircraft weapons in the Maus and Tiger II as follows:

Krupp presented a proposal for mounting the M.G.151/20 in each of the turrets for the Maus and Tiger with weapons mounts that could be elevated. The elevation arc is +20° to +90° in the Maus turret and -1° to +75°. In addition a sketch of a Maus turret is included with a fixed mounting for a weapon firing straight up. ***Wa Pruef 6*** *is of the opinion that the movable mount is most certainly more advantageous. An attempt should be made to mount the weapon so that it can be depressed to -8° for use in engaging ground targets.*

On 19 May 1943, Krupp sent a new drawing to **Wa Pruef 6** with the following message:

As requested, we are sending the enclosed drawing AKF 81292 B ***"Einbau des M.G.151/20 in Tiger"*** *(mounting the M.G.151/20 in the Tiger). The machinegun with a narrower gun shield is mounted in its production series fixed carriage in a cast armor piece on the turret. As a preliminary sighting arrangement, it is connected to the blade sight through a parallelogram rod, and the viewing port is covered by an armor glass block. Elevation is by hand; traverse by turning the entire turret. The elevation arc of the entire weapon runs from +12° to 75°. Increased depression is not planned because of the depth of the slit in the armor that would be needed for the gun barrel. The weapon is fired with an electrical circuit utilizing a magnet. The ammunition box holds a belt with 67 rounds.*

6.5.9.3 Tiger B mit 10.5 cm Kw.K. (L/68)

To start the process of up-gunning Panzers that would be needed for maintaining supremacy on future battlefields, Krupp in November 1944 created a series of conceptual design sketches with longer and heavier armament. On 20 January 1945, **Wa Pruef 6** sent to the **Generalinspekteur der Panzertruppen** a position paper stating their opinion on the proposals from Krupp for up-gunning the Panzers, including the **Tiger B mit 10.5 cm Kw.K. (L/68)**, as follows:

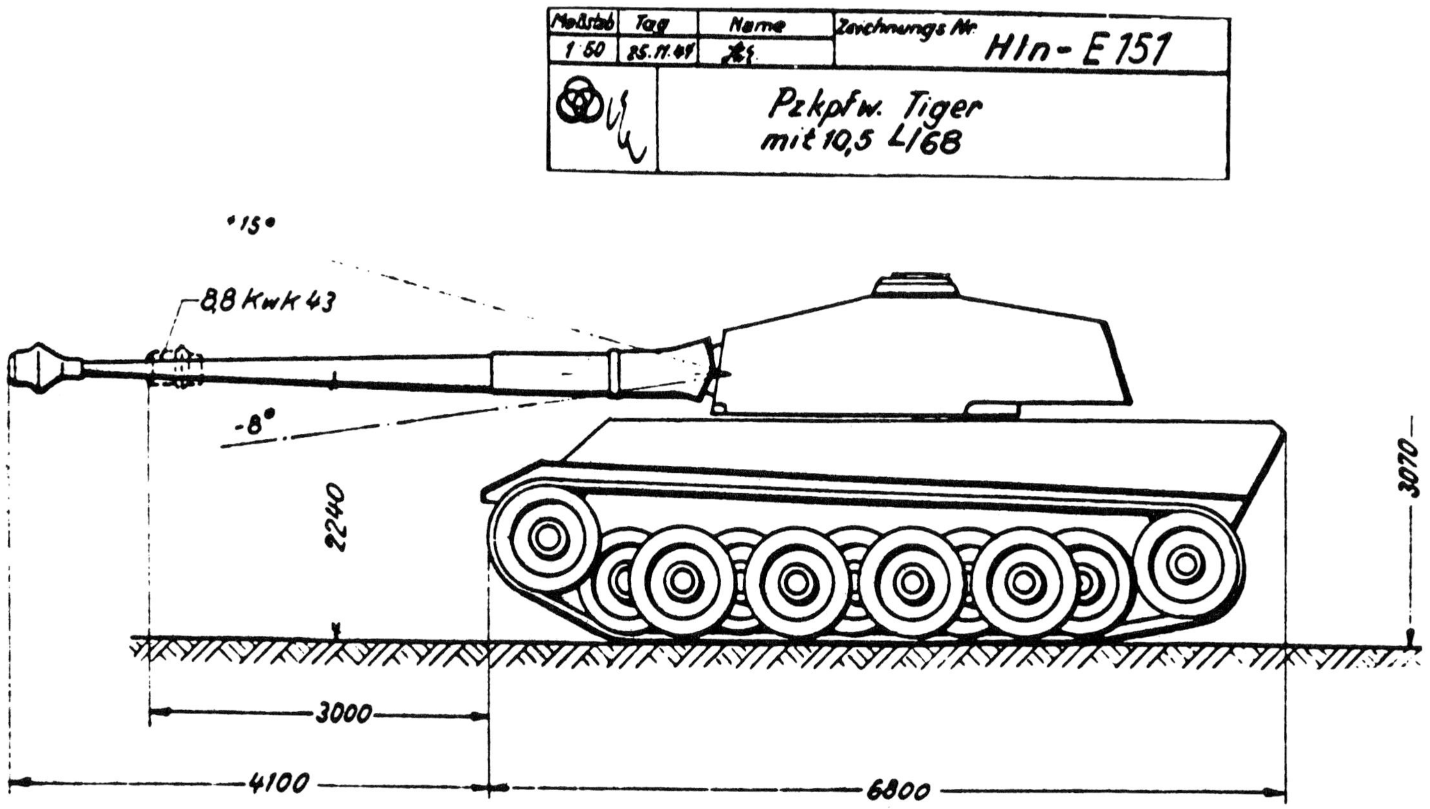

A sketch prepared by Krupp for a proposal to mount a **10.5 cm Kw.K. L/68** in the **Tiger II** turret.

*The proposed **10.5 cm Kanone** has not been accepted into the **Heer**. Therefore, it doesn't appear prudent to acquire a special caliber here. In any case, the gun would need new aiming devices and the turret would probably need to be redesigned.*

Because two piece ammunition would have to be used at first, the rate of fire would be significantly reduced. In addition, a second loader is needed, and we would run into significant difficulties finding space for him.

Appendices

APPENDIX A1 - PORSCHE "TIGER P2" PROPOSALS

Porsche Document Sk.7949 dated 5 October 1942

Model	180A	180B	181A	181B	181C
Normal Crew Access	Turret	=	=	=	=
Emergency Escape	Turret	=	=	=	=
Main Gun	8,8 L/71	=	=	=	=
Ammunition					
Length mm	1200	=	=	=	=
Weight Kg	23.5	=	=	=	=
Stored in Turret	16	=	=	=	=
in Hull	42	=	=	=	=
on Floor	10	=	=	=	=
Machineguns	1 MG42	=	=	=	=
Total Weight	64000	=	=	=	=
Track					
Pitch mm	130	=	=	=	=
Width mm	640	=	700	=	=
Length mm	4115	=	=	=	=
Ground Pressure Kg/cm^2	1.15	=	1.06	=	=
Suspension					
Model	101	180	=	=	=
Number of Springs	6	=	=	=	=
Pairs of Roadwheels	12	=	=	=	=
Weight/Roadwheel	4650	=	4620	=	=
Type of Drive Train	Electrical 2 Generators 2 Motors	=	Hydraulic Voith II		
Drive Sprocket Ø mm	794	=	=	=	=
Maximum Speed	35	=	=	=	=
Slope Climbing Ability	30°	=	=	=	=
Forward Idler					
Wheel Ø mm	794	=	=	=	=
Brakes	in Wheel	=	=	=	=
Track Adjustment	Inside	=	=	=	=
Stopping Brakes	Handwheel	=	=	=	=
Auxiliary Motor	T 141	=	=	=	=
Engine Cooling	Air	=	=	=	=
Engine Model	101/3	101/4	101/4	180/1	180/2
Type	Gasoline Porsche	= =	= =	Diesel Porsche Deutz	= = =
Fuel Injection	No	=	=	Yes	=
Number of Cylinders	10	=	=	16	=
Bore mm	115	=	=	110	135
Stroke mm	145	=	=	130	160
Volume liters	15	=	=	19.6	37
Speed rpm	2000	=	=	2000	=
Horsepower	300	=	=	370	700
Engines per Vehicle	2	=	=	=	1
Power Ratio HP/t	8.5	=	=	10.4	9.8

APPENDIX A2 - PORSCHE DESIGN DATA
Dated 21 October 1942

Porsche Typ	180	181
Characteristic	Petrol Electric	Diesel Hydraulic
Type	VK 4502 P	VK 4502 P
Overall Length		
Gun forward	10710 mm	=
Gun to the rear	9050 mm	=
Overall Width		
Combat Track	3320 mm	3400 mm
Loading Track	3140 mm	=
Overall Height	2950 mm	=
Length over Track	6670 mm	=
Hull Type of Cutting	Flame	=
Length, outside/inside	6860/5750 mm	=
Width, outside/inside	3180/1800 mm	=
Height, outside/inside	1457/1225 mm	=
Width, Basket	1570 mm	=
Greatest Bore Distance	6200 mm	=
Ground Clearance	480 mm	=
Gauge	2680 mm	2740 mm
Obstacle Climbility	800 mm	=
Gun Firing Height	2130 mm	=
Engine (Type)	Petrol Air cooled	Diesel Air cooled
Torque at Maximum	2 x 104 mkg	240 mkg
Power Output	2 x 350 HP	720 HP
Maximum Torque	2 x 115 mkg (n = 1900 rpm)	240 mkg (n = 1900 rpm)
Fuel Consumption	260 gm/HP	185 gm/HP
Cooling Fan (Type)	Axial (2/engine)	Radial (1 per cyl.)
Fan Power Consumption	2 x 35 HP	70 HP
Air Volume	2 x 5.0 m^3/sec 43200 kg/h	1 x 9.5 m^3/sec 41000 kg/h
Transmission (Type)	Petrol hydraulic	Diesel Hydraulic
Gear Change	Electric	Hydraulic
No. of Gears	Stepless	Pair convertors for steps
Total Reduction	about 1:110	about 1:150
No. of Main Gears	None	8 + 3
No. of Auxiliary Gears	None	4
Steering (Type)	Electric	Hydraulic-Mechanical
Intermediate Reduction	None	1:1
No. of Gears	None	22 spur gears
Final Drive (Type)	Epicyclic	Epicyclic and spur
No. of Gear Wheels	2 x 7	=
Total Gear Reduction in Bottom Gear	Generators in Series	Curves from Voith
Total Gear Reduction in Top Gear	Generators in Parallel	Curves from Voith
Total Number of Gear Wheels Engine to Track	2 x 7	
Total Number of Clutches Engine to Track	2 slipping clutches	
Drive (Type)	Rear	=
Driving Sprocket	Cast steel with steel gear rim	=
Pitch Circle	794 mm	=
No. of Teeth	19	=
Height/Width	60/40	=
Track (Type)	Dry	=
Track Support Length	4115 mm	=

Track Width	640 mm	700 mm
Track Pitch	130 mm	=
Top Rollers (Type)	Twin Rollers	=
Number	6 per side	=
Outer Diameter	700	=
Width of Rubber	4 inner rubber rings, each 31 mm	
Rubber Rings D:d	633/535 mm	=
Suspension (Type)	Porsche Toggle	=
Arm Bearing	Taper roller	=
Spring (Type)	Torsion bar	=
Armour Thicknesses		
Turret Front	100 mm	=
Turret Side	80 mm	=
Turret Top	40/25 mm	=
Turret Rear	80 mm	=
Hull Front	80 mm	=
Hull Side	80 mm	=
Hull Top	40/25 mm	=
Hull Rear	80 mm	=
Hull Belly	40/20 mm	=
Total Weight	65 tons	=
Power/Weight Ratio	10.8 HP/ton	11.1 HP/ton
Center of Gravity	About 120 mm in front of center with gun forward	
Ground Pressure	1.22 kg/mm^2	1.12 kg/mm^2
Track Length/Guage	1.53:1	1.50:1
Rubber Shear Stress	1.4 kg/cm^2	=
Max. Speed	About 35 km/h	=
Min. Speed	About 1.3 km/h	=
Max. Angle of Climb	35°	=
Fuel Tank Capacity	820 liters	=
Bogging Ability	1.7	=
Turning Radius	On the spot	=
Wading	Up to 6 m	=
Radius of Action at 45 kg/t Resistance	about 157 km	about 217 km
Radius of Action at 90 kg/t Resistance	about 72 km	about 99 km
Radius of Action at 350 kg/t Resistance	about 22 km	about 31 km
Ground pressure		
at 50 mm	1.11 kg/cm^2	1.04 kg/cm^2
at 100 mm	1.06 kg/cm^2	0.97 kg/cm^2
at 200 mm	0.96 kg/cm^2	0.88 kg/cm^2
Volume of Hull (Empty)	about 16.3 m^3	=
Volume of Fighting Compartment (Empty)	about 8.5 m^3	=
Volume of Fighing Compartment with Components	about 4.5 m^3	=
No. of Torsion Bars	6	6

APPENDIX B1- PANZERKAMPFWAGEN "TIGER II" DATA

Wa Pruef 6 dated 14 December 1943
(Current as of 1 November 1943)

1. Weights
   - Combat Weight .... 68,000 kg
   - Turret Weight .... 13,500 kg
2. Speeds
   - Highest Speed .... 42 km/hr
   - Sustained Speed on Roads (Autobahn) .... 38 km/hr
   - Sustained Speed on Average Terrain .... 15-20 km/hr
3. Range
   - Road .... 250 km
   - Average Terrain .... 145 km
4. Fuel Consumption per 100 km
   - Road .... 325 l
   - Average Terrain .... 575 l
5. Fuel Capacity .... 860 l
6. Measurements
   - Length Overall
   - with Gun Forward .... 10.286 m
   - with Gun Aft .... 9.966 m
   - without Gun Overhang .... 7.380 m
   - Gun Overhang with Gun Forward .... 2.906 m
   - Width Overall with Cross Country Tracks .... 3.580 m
   - Wheel Base .... 2.790 m
   - Track Width .... 0.800 m
   - Width Overall with Rail Loading Tracks .... 3.2780 m
   - Wheel Base .... 2.610 m
   - Track Width .... 0.660 m
   - Free Width Inside Hull .... 1.760 m
   - Free Diameter Inside Turret Ring .... 1.850 m
   - Height Overall .... 3.062 m
   - Turret Height with Commander's Cupola .... 1.217 m
   - Height of Centerline of Gun at 0° Elevation .... 2.245 m
   - Ground Clearance .... 0.500 m
   - Track Contact Length/Wheel Base 1.475
   - Ground Pressure .... 1.034 kg/cm^2
   - with Tracks Sinking in 20 cm .... 0.787 kg/cm^2
7. Capabilities
   - Engine Type .... Gasoline
   - Engine Make .... HL 230 P30
   - Power .... 700 PS
   - Power to Weight Ratio .... 10.3 HP/ton
   - Trench Crossing Ability .... 2.5 m
   - Step Climbing Ability .... 0.85 m
   - Grade Climbing Ability .... 35°
   - Fording Ability .... 1.75 m
   - Submerged Crossing Ability .... 4.0 m

8. Armor

Turret

Gun Mantlet	150 @ 13°
Front	110 @ 10°
after 51st Turret	180 @ 10°
Side	80 @ 21°
Rear	80 @ 20°
Roof	25/40/25 @ 50°
after 51st Turret	40

Hull and Superstructure

Driver's Front (Glacis)	150 @ 50°
Lower Hull Front	100 @ 50°
Side	80 @ 0-25°
Rear	80 @ 30°
Roof	40
Belly, Forward	40
Belly, Aft	25

9. Armament

1 - 8.8 cm Kw.K. L/71
1 - M.G. in turret, 1 M.G. in hull
1 - M.P.
1 - Smoke Discharger Device

10. Ammunition

A.P. and H.E. 66

11. Crew

5 men; 1 commander, 1 gunner, 1 loader, 1 driver, and 1 radio operator

12. Observation and Aiming Devices

Periscopes in cupola for commander
T.Z.F.9d gun sight for gunner
Periscope for Loader

13. Traverse and Elevation

+15 to -8° arc of elevation, 360° traverse

14. Direction Finding Device

Kurskriesel (Course Compass)

15. Radio Sets

Ultra-Shortwave Fu.2 receiver
Ultra-Shortwave Fu.5 receiver and transmitter

16. Other

Bordsprechanlage 1 (intercom)

APPENDIX B2- TIGER B TECHNICAL SPECIFICATIONS

by Henschel dated 18 September 1944 for the **Neue Turm-Ausfuehrung. Ab 48. Fahrzeug** (new turret model starting with the 48th vehicle).

General Vehicle Data
Combat Weight .... 69,800 kg *
Rail Loading Weight .... 66,300 kg *
Speeds
Sustained Speed
 on Roads (Autobahn) .... 38 km/hr
 on Average Terrain .... 15-20 km/hr
Range on Roads (Autobahn) .... 170 km
 on Average Terrain .... 120 km
Capabilities
Trench Crossing Ability .... 2.50 m
Step Climbing Ability .... 0.85 m
Slope Climbing Ability .... 35°
Fordability .... 1.60 m
Crew 5 men
Vehicle Measurements
Length Overall
 with Gun Forward .... 10,286 mm *
 with Gun Aft .... 9,966 mm *
 without Gun Overhang .... 7,380 mm
Gun Overhang with Gun Forward .... 2,906 mm *
Width Overall, over Track Guards .... 3,755 mm
Height Overall .... 3,090 mm
Length over the Tracks .... 6,400 mm
Width over the Tracks
 with the Cross Country Tracks .... 3,625 mm
 with the Rail Loading Tracks .... 3,270 mm
Ground Contact Length of the Tracks .... 4,120 mm
 by Sinking in 20 cm .... 5,400 mm
Width between Track Centers (Wheelbase)
 with the Combat Tracks .... 2,790 mm
 with the Rail Loading Tracks .... 2,610 mm
Height from Ground to Top of Chassis .... 1,860 mm
Firing Height of the Main Gun .... 2,260 mm
Ground Clearance
 Front .... 495 mm
 Rear .... 510 mm
Weight and Volumes
Weight of the Automotively Operational Chassis
 without the Turret, Weapons, Ammunition,
 and Equipment .... 52,000 kg
Total Volume .... 17.4 m^3
Fighting Compartment Volume .... 11.0 m^3
Engine Compartment Volume .... 2.2 m^3
Turret Volume above the Deck .... 4.2 m^3
Volume of the Rest, outside of the Engine
 and Fighting Compartments .... 2.0 m^3
Hull
Bare Weight .... 28,000 kg
Empty Weight after Machining .... 27,700 kg
Largest Outside Length of the Hull .... 7,134 mm
Hull Width Outside at Suspension .... 1,928 mm
 at the Height of the Panniers (Middle) .... 2,938 mm
Hull Width Inside at Suspension .... 1,768 mm
 at the Height of the Panniers (Middle) .... 2,778 mm
Height of the Hull from Belly to the Deck
 Forward .... 1,365 mm
 Rear .... 1,350 mm

Hull Armor	Thickness at Angle	
Driver's Front Plate (Glacis)	150 mm at 50°	
Hull Front Lower	100 mm at 50°	
Hull Side Upper	80 mm at 25°	
Hull Side Lower	80 mm at 0°	
Deck	40 mm at 90°	
Belly	40 mm at front, 25 mm at rear at 90°	
Suspension		
	Cross-Country Track	Loading Track
Number of Guides per Track Link	2	2
Track Pin Length	818 mm	658.5 mm
Track Pin Diameter	24 mm	24 mm
Ground Pressure without Sinking In	1.02 kg/cm^2	1.23 kg/cm^2
Sinking In 20 cm	0.777 kg/cm^2	0.943 kg/cm^2
Steering Ratio (Track Contact Length divided by Wheel Base)	1.475	1.158
Roadwheels		
Type of Suspension	**Gestaffelt** (overlapping)	
Type of Roadwheels	Steel tires with rubber cushioning	
Number of roadwheels per Side	9	
Roadwheel Diameter	800 mm	
Weight Carried per Roadwheel	3,610 kg	
Axle Distance from Roadwheel to Roadwheel	515 mm	
Play between Roadwheel and Track Guide	2 mm	4 mm
Suspension Arms and Suspension		
Material for Bushings	Novetext	
Type of Suspension	Torsion Bars	
Number of Torsion Bars	18	
Torsion Bar Diameter	60 or 63 mm	
Torsion Bar Head Diameter	90 mm	
Active Torsion Bar Length	1,800 mm	
Total Torsion Bar Length	1,960 mm	
Distance from Hull Bottom to Center of Bar	95 mm	
Drive Wheel, Idler Wheel, and Shock Absorber		
Effective Diameter of the Drive Wheel	870 mm	
Segment Spacing for the Drive Wheel	151 mm	
Diameter of the Idler Wheel	650 mm	
Largest Distance for Track Adjustment	210 mm	
Number of Shock Absorbers	4	
Automotive System		
Engine Designer	Maybach-Motorenbau	
Type HL 230 P30		
Power at 3000 rpm, 20°C air temperature, and 760 mm air pressure	750 HP	
Engine Weight	1,300 kg	
Number of Cylinders	12	
Stroke	145 mm	
Bore	130 mm	
Swept Volume	23,000 cm^3	
Compression Ratio	1 to 6.8	
Ignition	2 Magnetos with built in **Zuendfunkschnapper**	
Fuel Consumption per 100 kilometers		
on Roads	500 l	
on Average Terrain	700 l	

Engine Cooling System
Type of Cooling .... Water Cooled
Number of Radiators 4
Radiator Block Height .... 324 mm
Radiator Block Width .... 522 mm
Radiator Block Depth .... 200 mm
Area of Radiator Face .... 0.169 m^2
Number of Fans .... 2 double fans
Diameter of the Fan Wheel .... 520 mm
Speed of the Fans at Maximum Engine Speed
in the Summer .... 3,765 rpm
in the Winter .... 2,680 rpm
Type of Fan Drive .... Universal Gears with Connecting Shaft
Maximum Power Needed for Fans .... 40 HP
Number of Air Filters 2
Manufacturers of Air Filters .... Mann & Hummel
Transmission and Steering Gears
Length of Connecting Drive Shafts
Front .... 993 mm
Rear .... 1,187 mm
Transmission Type .... Maybach OG 40 12 16 B
Manufacturers .... Maybach and Zahnradfabrik
Number of Forward Gears 8
Number of Reverse Gears 4
Overall Length of the Transmission .... 1,266 mm
Overall Width of the Transmission .... 600 mm
Overall Height of the Transmission .... 620 mm
Vehicle Speed with Engine Speed at 3000 rpm
1.Gear .... 2.57 km/hr
2.Gear .... 3.83 km/hr
3.Gear .... 5.62 km/hr
4.Gear .... 8.33 km/hr
5.Gear .... 12.75 km/hr
6.Gear .... 18.95 km/hr
7.Gear .... 27.32 km/hr
8.Gear .... 41.5 km/hr
Reverse .... 3.39 km/hr
Overall Gear Reduction of the Transmission .... 1 to 16
Gear Cutting in the Universal Gear .... Spiral
Gear Reduction in the Universal Gear .... 1 to 1.05
Steering Gear Manufacturer .... Henschel & Sohn
Steering Gear Type .... Double Radius L801
Gear Reduction of the Steering Gear .... 1 to 1.2955
Number of Steering Stages 2
Number of Steering Clutches 4
Material for Clutch Facing .... Jurid or Emero
Number of Gears 25
Minimum Turning Radius .... 2.08 m
Largest Turning Radius .... 114 m
Steering Controls .... Argus Steering Device
Total Length of Transmission
with Steering Gear .... 1,690 mm
for 12 trial model L801 .... 1,790 mm
Total Weight of Transmission with Steering Gear .... 1,200 kg

<u>Final Drive and Brakes</u>
Gear Reduction of the Final Drive .... 1 to 12.56
Type of Brakes .... Argus Disk LB 900.4
Outer Diameter of the Brake Drum .... 565 mm
Material for the Brake Lining .... Ge.
Brake Cooling .... Cooling Ribs on Housing
Type of Brake Operation .... Mechanical Foot and Hand Levers
<u>Volumes</u>
Fuel Capacity .... 860 l
<u>Traversing Turret</u>
Turret Weight .... 13,500 kg
Empty Weight of the Machined Turret Shell .... 8,000 kg
Turret Height with Commander's Cupola .... 1,217 mm
Free Inner Turret Ring Diameter .... 1,850 mm
<u>Armament</u>
8.8 cm Kw.K.43 (L/71)
Ammunition (68 stowed, 16 loose on turret platform) 84
M.G.42 (in the interim M.G.34) (1 anti-aircraft) 3
32 Sacks with 150 Rounds of Belted Ammunition 5,850
M.P. 1
6 Magazines, each with 32 Rounds 192
Explosive Charges 3

The data marked with an "*" are different for **Fahrzeugen 1 bis 47** (series production chassis numbered 1 through 47) with the older **Turm-Ausfuehrung (Porsche)** (turret designed for the Porsche model) as follows:
Combat Weight .... 68,500 kg
Rail Loading Weight .... 65,000 kg
Length Overall ....
with Gun Forward .... 10,280 mm
with Gun Aft .... 9,860 mm
Gun Overhang with Gun Forward .... 2,900 mm

APPENDIX C - PART NUMBERS FOR TURRETS

Parts List	**021St48010** for **VK 45.02(P)** electric traverse	**021St48010** for **VK 45.03** hydraulic traverse	**021St50600** for **VK 45.03** hydraulic traverse
Dated	21Oct42	1Feb44	1Jul44
Turmgehaeuse (Armor Body)	48011	=	50601
Turmkugellager (Ball Bearing Race)	48012	=	50602
Kommandantenkuppel (Commander's Cupola)	48013	=	2762
Wiegenpanzer (Gun Mantlet)	48014	=	50604
Wiegenabdichtung (Seal for Gun Mantlet)	48015	=	N/A
Lukendeckel, oberer (Loader's Hatch)	48016	=	50606
Lukendeckel, hinterer (Rear Hatch)	48017	=	50607
Schaulochdeckel (View Port Cover)	48018	Welded Shut	N/A
Turmschwenkwerk (Traverse Gear)	48019	50608	50608
Schwenkwerktraeger (Traverse Gear Support)	48020	50609	50609
Drehbuehne (Turret Platform)	48021	=	50610
Hohenrichtmaschine (Elevation Gear)	48022	48023	=
M.G.-Lagerung (Machinegun Mount)	48024	=	50612
M.G.-Abzug (MG Firing Mechanism)	48025	50613	50613
Abdichtung M.G.-Lagerung (Seal for MG Mount)	48026	=	N/A
Lagerung fuer Optik (Gun Sight Mount)	48027	=	50615
Munitionszufuehrung (MG Ammo Feed)		48028	50614
Turmzurrung (Traverse Lock)	48028	50620	50620
12-Uhrzeigerantrieb (Azimuth Indicator Drive)	48029	=	N/A
Zwoelfuhrzeiger (Azimuth Indicator)	48030	=	50617
MP-Stopfen (Pistol Ports)	48031	=	50618
Munitionslagerung (Ammunition Stowage)	48032	48036	50619
Huelsenanschlag (Cartridge Stop)	48033		
Steuerung z. FLuessigkeitsgetriebe (Control for Hydraulic Drive)		50621	50621
Zwoelfuhrzeiger (Modified Azimuth Indicator)			50622

Lagerung fuer Flieger-MG (Mount for Anti-Aircraft MG)	48037		
Kommandantensitz (Commander's Seat)	48040	50625	50625
Richtschuetzensitz (Gunner's Seat)	48041	50626	50626
Ladeschuetzensitz (Loader's Seat)	48042	3AKF31785	50627
Zubehoerlagerung (Equipment Stowage)	48044		
Elektrische Einrichtung (Electrical Equipment)	48045	2794	50630
Umformerschutzkastens (Transformer Box)	48047	N/A	
Schleifringuebertrager (Slip Ring Contacts)	48048		
Fahrtregler (Drive Control)	48049	N/A	
Nebelwurfgeraet, Links (Smoke Candle Discharger)	41406	N/A	
Nebelwurfgeraet, Rechts (Smoke Candle Discharger)	41407	N/A	
Kompressoranlage (Compressor)		50638	50638
Turmlukendeckel, oberer (Modified Loader's Hatch)			50640
Pz-Fuehrerkuppel (Modified Commander's Cupola)			50641

APPENDIX D - ARMOR SPECIFICATIONS

ARMOR FOR TURM NR.1-50

The armor plate on the first 50 turrets for the Tiger II (**Turm Nr.1-50**) had been manufactured much earlier (in 1942) than the armor plate for the **Serienturm** and therefore under a different specification. Armor plate for the front and sides of these 50 turrets was fabricated by Krupp in accordance with their specifications for **PP793** and **PP7182** rolled armor plate. The tensile strength specified for 55 to 80 mm thick plates was 100 to 115 kg/mm^2 (equivalent to Brinell Hardness Number range of 294 to 338) and for the 85 to 120 mm thick plates was 95 to 105 kg/mm^2 (equivalent to Brinell Hardness Number of 279 to 309). A decision had been made by May 1942 to harden the face of the armor to increase resistance to penetration. The specification for the hardened face was set at 200 kg/mm^2 (equivalent to Brinell Hardness Number of 555) to a depth of 4 to 6 mm. The hardness specification for the interior and back of face-hardened plates remained unchanged from that specified for homogeneous armor plate.

The composition of **PP793** armor steel alloy was 0.34% carbon, 0.42% manganese, 0.39% silicon, 2.32% chromium, and 0.22% molybdenum. A decision was made in May 1942 to replace **PP793** with a richer alloy, named **PP7182**. **PP7182** was to have the same composition of carbon, manganese, silicon, and molybdenum as **PP793**, but with slightly less chromium (1.7 to 2.3%) and enriched with scarce nickel (0.40 to 1.00%). The combined percentage of chromium and nickel in **PP7182** was to be at least 2.7%. A list of turrets reveals that **PP793** was used in turret numbers as high as 150116 and **PP7182** was used in turret numbers as low as 150111.

After capture a **Turm Nr.1-50** was removed from a Tiger II and shipped to England where it was tested. The following information was extracted from the Preliminary Report on Armour Quality and Vulnerability of "Tiger" II Turret, examined at Chobham on 21 and 22 September 1944:

Plate	Thickness	Angle	Corrected Poldi Hardness Outer Surface	Inner Surface
Front	100 mm	Curved	472	280
Right Side	82 mm	30°	535-565	325
Left Side	80-88 mm	30°	305-330	296
Rear	80	30°		
Roof Front	40	80°	278	—
Roof Rear	40	80°	300	—

The D.T.D. thickness figures were checked by drilling holes in the armor, and the angles were established independently by measurements with a clinometer and by a complete scale drawing. The bulge underneath the commander's cupola on the left side was 88 mm thick. The top part of the curved turret front plate was sloped at an angle of 40°. The only part of the turret front that was reported as vulnerable to 6 pdr. A.P.C.B.C. attack were areas on either side of the gun mantlet, each about 12 inches wide by 8 inches tall. The total frontal area vulnerable to the 17 pdr. A.P.C.B.C. at short range was only about 4 feet by 2 feet, most of which constituted a difficult target which could be perforated only at ranges of less than 1000 yards from directly ahead.

With the following two exceptions, the British tests confirmed compliance with the basic armor specifications under which these turrets were to be fabricated: 1) the left side of the turret, shaped by "pressing", was not face hardened, and 2) the bulge underneath the cupola, created during pressing, was thicker than the rest of the plate. The original Krupp drawing shows the bulge under the cupola as having the same thickness as the rest of the turret side. The Poldi portable hardness tester used by the British to test the surface hardness is not the most accurate method for measuring face-hardened plate. But this type of measurement was sufficiently accurate to spot when it was or wasn't face-hardened.

ARMOR FOR VK 45.03 WANNE UND SERIEN-TURM

A standardized (**Einheits)** specification was in effect when most of the armor plate was created for the hulls and **Serien-Turm** for the Tiger II. Thinner 25 mm plates were made from **E32** armor steel, the 40 mm and 80 mm plates from **E22** armor steel, and the thicker front plates from **E40**, **E41**, and **E43** armor steel.

Specifications for these standardized armor plates were:

Armor Type	Spec. Date	Thickness	Resistance Kg/mm^2	Brinell Hardness	Alloy Percent
E32	Dec42	16-30 mm	105-120	324-370	3.1-4.2
E22	Feb43	35-50 mm	95-110	279-324	2.2-2.8
E22	Feb43	55-80 mm	90-105	266-311	2.2-2.8
E22	Jun44	85-120 mm	75-90	220-265	2.2-2.8
E40	Mar44	125-160 mm	75-90	220-265	2.45-3.35
E41	Jun44	125-160 mm	75-90	220-265	3.9-5.1
E43	Jun44	165-200 mm	75-90	220-265	4.3-5.2

Composition of each of the standardized armor plates was a bit different with a higher alloy percentage used in thicker plates and scarce nickel used in the thicker frontal armor plates, as shown in the following table:

Armor Steel Alloy Composition in Percentage

	C	Mn	Si	Cr	Ni	Mo	V	P&S
E22	0.37-0.47	0.60-0.90	0.20-0.50	1.60-1.90	—	—	<0.15	<0.05
E32	0.37-0.47	0.60-0.90	0.20-0.50	1.20-1.60	1.30-1.70	—	<0.15	<0.05
E40	0.32-0.42	0.30-0.65	0.15-0.50	2.00-2.40	—	0.15-0.30	—	<0.05
E41	0.30-0.40	0.60-0.90	0.20-0.50	2.30-2.70	1.00-1.50	—	<0.15	<0.05
E43	0.28-0.33	0.60-0.90	0.20-0.50	2.80-3.20	0.90-1.10	—	—	<0.05

There is no proof that substandard German armor plate was used during the last years of the war. All original documents confirm compliance with standard specifications throughout the war. Within the guidelines of these specifications, armor suppliers were authorized to request approval for acceptance of minor deviations outside the range of chemical composition or hardness. These requests for exceptions, reviewed on a case-by-case basis, were frequently rejected. Standards for testing resistance to penetration and mode of failure were conducted by firing armor-piercing projectiles at representative test samples from each melt. Procedures were also established on the course of action to take if the armor plates failed to meet the specifications. The **Waffenamt** acceptance inspector at Krupp, Essen reported on 5 May 1944: *Test firing has been conducted for **Schmelze 594** (melt number 594) and it is released. **Schmelze 414** has not passed the test firing. Plates fabricated from this melt are to be thrown away.* Again on 10 May 1944, the inspector reported: ***Schmelzen 439 und 9553** didn't pass test firing, and the 10 rolled plates for Tiger II turrets, created from these melts, are to be thrown away.*

British examination of a captured Tiger II (**Fgst.Nr.280093** or **280103**) in Report No.17 by the A.F.V.(T) 21 Army Group, stated: *There is no evidence of face hardened armour, although it was found on some plates of the prototype [sic] turret. Some surface hardness figures have been taken by means of the "Poldi" Portable Brinell equipment and these are as follows:*

Turret Front Plate (180 mm at 10°) 190 - 200 Brinell
Glacis Plate (150 mm at 50°) 210 - 220 Brinell
Turret Side Plate (80 mm at 20°) 230 - 240 Brinell
Pannier Side Plate (80 mm at 25°) 260 - 280 Brinell

It should be noted that these figures have been taken at the surface of the plate without grinding away any of the armour surface; accordingly then, it is possible that the hardness figures are rather low as there is probably some degree of surface decarburization.

In the Notes of Enemy A.F.Vs. Pz.Kpfw.Tiger. Model B. ('Royal Tiger') - Sd.Kfz.182 from M.I.10 dated May 1945, data on the armor were reported as:

Plate	Thickness in mm	Angle to Vertical in degrees	Approx.Brinell Hardness figure
Front glacis plate	150	50	220
Front nose plate	100	50	220
Side superstructure	80	25	275
Pannier floor	25	horizontal	
Side hull plates	80	vertical	264
Top superstructure	40	horizontal	345
Belly plate, front	40	horizontal	
Belly plate, rear	25	horizontal	
Tail plate	80	30	225
Turret, top front	44	80	325
Turret, top center	44	horizontal	325
Turret, top rear	44	80	325
Turret, side	80	20	290
Turret, rear	80	20	290
Turret, front	185	10	220

Sources and method of measurements were not reported.

Also from the Publisher

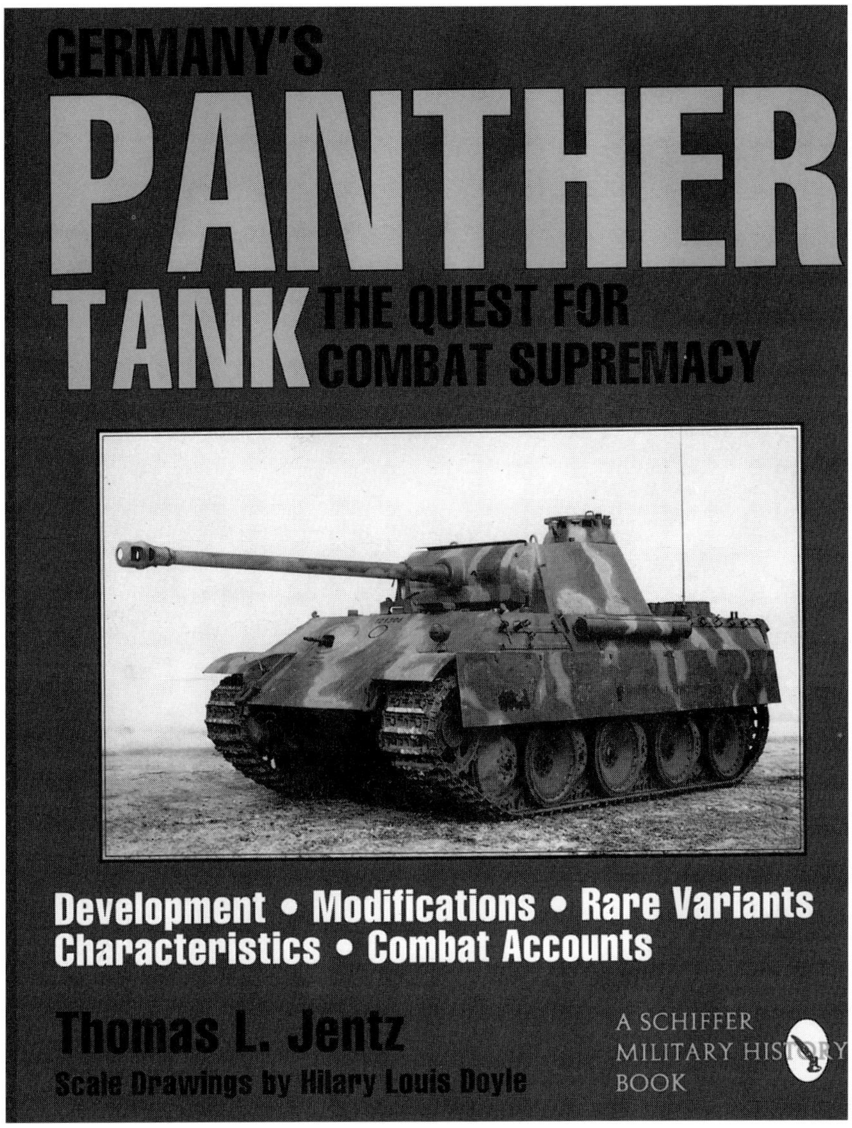

GERMANY'S PANTHER TANK: THE QUEST FOR COMBAT SUPREMACY
Development • Modifications • Rare Variants Characteristics • Combat Accounts

Thomas L. Jentz

Over twenty years of research went into the creation of this history of the development, characteristics, and tactical capabilities of the Panther. New information was found on the evolution of the Panther's predecessors, the significant modifications made during the production run of each Ausf., how the Panther II fit into the series, the Panthers tactical characteristics and capabilities, and experience reports written by those who fought with and against the Panther.
Size: 8 1/2" x 11" over 170 b/w photos, line drawings, charts, appendices
160 pages, hard cover
ISBN: 0-88740-812-5 $45.00

PANZERTRUPPEN
The Complete Guide to the Creation & Combat Employment of Germany's Tank Force
1933-1942

Edited by Thomas L. Jentz

In September of 1939, the world was astounded by Germany's ability to defeat Poland in less than a month. With the world still puzzled by the suddenness of this event, Denmark, Norway, Holland, Belgium, and France fell in rapid succession to the German onslaught, leaving Britain in shock. Greece and Yugoslavia were rapidly overrun during April of 1941, while German-Italian forces advanced rapidly in North Africa. Russia's turn was next, when German forces began pulverizing their forces in June of 1941. How had Germany achieved victory after victory, often against numerically superior enemy forces? The answer came in two words – Panzer and Blitzkrieg. When and how had Germany built its Panzer forces and trained them for the Blitzkrieg? When was each Panzer unit formed? What was their organization? Why were Panzer units disbursed among the Panzer-Divisions, leichte Divisions, and Armee-Korps? When were the various types of Panzers developed? What were their armament, armor protection, and capability? How many of each type were produced? What tactics did they use? How successful were they in combat? This is the only book that provides detailed answers to these and other questions related to how German tankers trained and fought from 1933-1942.
Size: 8 1/2" x 11" over 100 b/w photographs, maps, charts, appendices
288 pages, hard cover
ISBN: 0-88740-915-6 $49.95

PANZERTRUPPEN
The Complete Guide to the Creation & Combat Employment of Germany's Tank Force
1943-1945

Edited by Thomas L. Jentz

Having failed to win their strategic objectives, the Heeres (German Army) was overextended deep in enemy territory facing numerically superior forces in the Fall of 1942. Now it was the Allies turn to launch the offensives that would keep the Heeres off balance. With few exceptions the days of employing massed Panzer formations to gain strategically decisive objectives had come to an end. And, these few exceptions were almost invariably failures. Forced onto the defensive the Panzertruppen modified their tactics becoming expert at counterattacks that spoiled their opponents drives and inflicted heavy casualties. They became so proficient that the Allied armies took years to retake territory that the Heeres had overran in a few weeks or months. This companion volume to *Panzertruppen 1933-1942* presents how the Panzertruppen fought during their defensive struggle with details on the units, organizations, types of Panzers, and tactics. The first volume presents the offensive phase up to October 1942.
Size: 8 1/2" x 11" 47 b/w photographs, maps, charts, appendices
304 pages, hard cover
ISBN: 0-7643-0080-6 $49.95

Authors Tom Jentz (right) and Hilary Doyle.